Feminist Bioethics in Space

Feminist Bioethics in Space

Gender Inequality in Space Exploration

Konrad Szocik

OXFORD
UNIVERSITY PRESS

OXFORD
UNIVERSITY PRESS

Oxford University Press is a department of the University of Oxford.
It furthers the University's objective of excellence in research, scholarship,
and education by publishing worldwide. Oxford is a registered trade mark of
Oxford University Press in the UK and in certain other countries.

Published in the United States of America by Oxford University Press
198 Madison Avenue, New York, NY 10016, United States of America.

Library of Congress Cataloging-in-Publication Data
Names: Szocik, Konrad, author.
Title: Feminist bioethics in space : gender inequality in space exploration / Konrad Szocik.
Description: New York, NY : Oxford University Press, [2024] |
Includes index.
Identifiers: LCCN 2024024071 (print) | LCCN 2024024072 (ebook) |
ISBN 9780197691045 (hardback) | ISBN 9780197691052 (epub)
Subjects: LCSH: Outer space—Exploration—Moral and ethical aspects. |
Feminist bioethics. | Outer space—Exploration—Philosophy. | Women in science. |
Sex discrimination in science.
Classification: LCC QB500.262 .S96 2024 (print) | LCC QB500.262 (ebook) |
DDC 174/.20820919—dc23/eng/20240617
LC record available at https://lccn.loc.gov/2024024071
LC ebook record available at https://lccn.loc.gov/2024024072

DOI: 10.1093/9780197691076.001.0001

Printed by Integrated Books International, United States of America

MIX
Paper
FSC   FSC® C183721

For Rakhat Abylkasymova, my wife and my only inspiration

Contents

Contents

Acknowledgments

I began work on this monograph at Yale University while completing my research appointment at the Yale Interdisciplinary Center for Bioethics. My sincere thanks go to Stephen Latham and Lori Bruce from the Center for allowing me to carry out my research there. The climate of feminism and equality that pervades Yale University created an ideal environment for my interests in feminism, bioethics, and the cosmos. I am also very grateful to Yale University Library for the excellent working conditions for this monograph. The work on the book was finalized at my home university, the University of Information Technology and Management in Rzeszow, Poland.

Many thanks to everyone who shared their comments and feedback on my book: Tony Milligan, Erika Nesvold, Karen Lebacqz, Lisa Stenmark, Evie Kendal, Koji Tachibana, Linda Billings, Gonzalo Munevar, and Rakhat Abylkasymova. My thanks also to the anonymous reviewers, as well as the entire Oxford University Press team involved in the production of my book, led by Jeremy Lewis and Madeline Hoverkamp. The book was supported by a grant from the National Science Center in Poland (UMO 2021/41/B/HS1/00223).

Special thanks go to my wife, Rakhat Abylkasymova, for her support and patience, as well as her extremely valuable comments and advice.

Introduction

Feminist bioethics of space exploration is a combination of words that we may look for in vain in the philosophical literature, as well as, more broadly, in the humanities and social sciences. Moreover, the bioethics of space exploration itself is a novel area and to date has only lived to see one monograph (Szocik 2023), while the combination of feminism and space exploration is unprecedented. It is noteworthy that in 2023, monographs began to appear raising feminist issues in the context of space exploration, albeit, with few exceptions (Kendal 2023),[1] not in relation to bioethical issues. One of them is the work of Erika Nesvold (2023), in which the author highlights the enrichment of the discussion of the future of humanity in space with a humanistic element, which, as Nesvold points out, is definitely lacking in the approach of those in the space sector.[2]

The purpose of this monograph is to fill this niche in the philosophy and bioethics of space exploration and, more broadly, in humanistic thinking about the future of humans in space. I propose a feminist perspective on potential selected problems in space such as human enhancement, gene editing, and reproduction. But, as I emphasize in the book, feminism is inherently an all-encompassing philosophical approach. Hence, the reader of this book will also encounter considerations that go beyond the scope of bioethics and take us into areas such as the very meaning of carrying out space missions and their potential consequences, as well as the exclusion of numerous groups of people on Earth. Such exclusion and discrimination—not only against women, but also of people of a different skin color, background, social class, or ability than the privileged group, and therefore also of many men—cast a shadow over future space policy, which is unlikely to be one of equality, justice, and inclusion. Although the bioethics of space missions considered from a feminist perspective is the focus of this monograph, it is impossible not to highlight and discuss other related elements that, according to feminist philosophy, cannot but affect the moral evaluation of bioethics in space.

The book contains six chapters. The first chapter provides an introduction to feminist ethics and bioethics. Feminism is almost absent from reflections on space missions and, in particular, bioethical challenges in space. Such an

Feminist Bioethics in Space. Konrad Szocik, Oxford University Press. © Oxford University Press 2024.
DOI: 10.1093/9780197691076.003.0001

introduction seems particularly valuable for all those who, even if they are familiar with the main ideas of feminism, would nevertheless like to learn more about the complexity of feminist philosophy, its originality, and its distinctiveness from mainstream thought.

Chapter 2 is an introduction to the feminist bioethics of space exploration, as well as both the methodology used in this book and the proposed methodology of feminist bioethics of space missions. The differences between feminist and nonfeminist approaches to the bioethics of space missions are discussed, and selected issues concerning the main bioethical principles, but related to space and considered from a feminist perspective, are analyzed. The main characteristics of the methodology inherent in feminist bioethics are also discussed.

Chapter 3 focuses on the issue of exclusion. This is one of the most important feminist categories, and an interest in exclusion as a structural problem is a hallmark of feminism. The chapter looks at problems of power structures, characterizes gender issues in the space environment, and discusses the exclusion of various social groups from participating in the exploration and exploitation of, as well as possible future settlement in, space.

The fourth chapter is devoted to disabilities in space. It discusses the issue of disability as understood by feminism and, in particular, in relation to such a specific environment as that of space. One of the issues considered is whether one can conceive of a justification for excluding people with disabilities from participating in space exploration. Another issue under discussion is whether it is morally possible to justify a preference for people without disabilities in the case of hypothetical reproduction in space, where we could decide the genetic condition of future children.

Chapter 5 examines the issue of human enhancement in light of feminist bioethics. Here we point out the ambiguity of the feminist assessment of human enhancement. What is more unequivocally shared by feminists is, in turn, a critique of the idea of the norm, as well as an unmasking of the social and ideological construction of science. Both issues are significant for the idea of human enhancement. The chapter contains a feminist critique of both somatic and germline gene editing. The case of moral bioenhancement is also discussed.

The sixth and final chapter is devoted to antinatalism and environmental issues. The idea of antinatalism, or the abandonment of procreation, was an ally of early feminist bioethics in feminists' struggle against the oppression of pronatalism. The chapter examines feminist arguments for and against antinatalism, both in the terrestrial and the cosmic contexts, especially hypothetical space settlement. Environmental issues related to climate change are also discussed in order to place such a grand and often idealized project as the future expansion of humanity in space in the context of exclusion and oppression experienced by a huge part of humanity.

Notes

1. Evie Kendal's (2023) chapter addresses issues inherent in the feminist bioethics of space exploration but is just one chapter on this topic in the entire collection (Schwartz et al. 2023).
2. In her book, Nesvold never once uses the word "feminism," but the nature of her deliberations carries a feminist tinge. Brian Patrick Green (2021) does mention feminism as one of the currents in ethics, but does not give a single example of the application of feminism to the subject of space exploration.

1

An Introduction to Feminist Ethics and Bioethics

Introduction: The Inferior Position of Women

The essence of feminist philosophy is the assumption that the situation of women is generally worse than that of men, and that this condition is something unjust, and not a desirable or inevitable state (Cameron 2020, 8). This last component, the recognition of the injustice of this inequality, is a novelty offered by feminism. For both in the past and today, many gender inequalities have been justified by referring to biology, psychology, culture, or upbringing. Paradoxically, inequality for, and discrimination against, women is still widely accepted as something if not just, then at least morally neutral. Feminism as both a philosophy and a movement of political action unequivocally exposes the injustice of this exclusion (Delap 2020).

Therefore, the axis of feminist reflection is to appeal for equal consideration of the interests of men and women. The feminist perspective is characterized by this search for manifestations of inequality between the interests of men and women (Purdy 1996, 5, 7). This classical understanding of feminism growing out of historical, structural injustice and inequality against women, however, should not obscure the fact that feminism expresses a theoretical and practical concern for justice and equality for all.[1] The essence of this expanded understanding of feminism that goes beyond the category of gender, humanist feminism, is illustrated by Mari Mikkola:

> As I see it, feminism is for everyone, not just for women. . . . In this sense, feminism is not just about women; it is about all individuals facing dehumanizing treatments due to sexist and/or patriarchal injustices. So the humanist feminism on offer does not reject the basic tenet of feminism that women as a group are subject to systematic and nonaccidental forms of injustice that are due to sexism and patriarchy. It simply highlights that women are not alone in having their lives so infringed. This being the case, humanist feminism tells us that sexist and patriarchal social

Feminist Bioethics in Space. Konrad Szocik, Oxford University Press. © Oxford University Press 2024.
DOI: 10.1093/9780197691076.003.0002

injustices do not pose problems just for women—rather, they pose problems for humanity. (Mikkola 2016, 267)

Third-wave feminists often emphasize that as a result of globalization, among other things, there has been a de-authorization and decentralization of patriarchy.[2] The position of top-down patriarchy has been replaced by the accumulation of power and capital by global corporations, which, perhaps somewhat paradoxically from the point of view of second-wave feminism (Freeman 1979), have denied access to power and large capital not only to women but also to huge numbers of men outside the small group with power.[3] The global economy of twenty-first-century capitalism has opened up equal opportunities for women and men in many spheres, as well as posing comparable challenges for both women and men, causing women of third-wave feminism to be more likely to be able to identify not only with women but also with men of their generation as facing the same challenges and having the same opportunities (Heywood and Drake 1997; 2004, 16, 18).

But despite this commonality of gender in oppression of an economic nature, caused by the dynamics of capitalism, women are often disadvantaged simply by being women, by their reproductive biology being used against them, and by entrenched social structures and stereotypical notions of gender.[4] The tendency to control women's reproduction is still so strong that the issue of restricting abortion is constantly recurring in democratic countries. The United States and Poland are recent examples where the patriarchal drive to oppress women by depriving them of their reproductive rights is still being implemented today.[5] As I argue in chapters 5 and 6, the space mission environment can foster a resurgence of anti-abortion sexist policies and a radical erosion of women's reproductive rights.

Like bioethics itself, feminism is not uniform. Moreover, neither are the philosophy, ethics, and bioethics of space missions. The differences between the various positions in the philosophy of space exploration concern the justification for particular types of missions, the dispute over the value of scientific exploration of space, investment in human or robotic missions, the moral status of space as an object of potential exploration, and the ethical assessment of the justification for human enhancement for space missions. The diversity within feminism does not particularly affect the nature of the considerations and arguments presented in this book, primarily because feminism itself is so radical and innovative compared to nonfeminist approaches to bioethics and, on the other hand, space mission ethics and bioethics do not discuss the feminist perspective that what is relevant here is simply to propose an application of feminism to space mission bioethics. For only feminism, with all its diversity, is interested in exposing the oppression and exploitation of marginalized groups at its starting point.

Thus, in the context of the fact that feminism is almost absent from the philosophy, bioethics, and politics of space exploration, as well as the fact that the space exploration environment is masculinized and associated with what is nonfeminist, namely nationalism, macho worship, and militarism, applying feminism to such a theoretical and practical environment seems to be a desirable value. Nevertheless, for some of the content discussed in this book, such as reproductive issues, including antinatalism, as well as human enhancement, there are interesting and important differences between various feminist positions. But even in such disagreements within the feminist bioethics of space exploration, I am inclined to minimize the significance of these differences as long as they grow out of a concern for women's rights, feelings, intentions, and interests. And these in the nonfeminist philosophy and bioethics of space exploration are generally not taken into account.

The different approaches within feminism expose the role played by each category. For example, liberal feminism emphasizes the importance of choice, of which women are often deprived. Radical feminism stresses the importance of control and emphasizes that women are often deprived of real control over their lives and their choices are ostensible, remaining in practice controlled by men or subordinate to men's choices. Cultural feminism, on the other hand, exposes the importance of caring and connections, which are context-dependent and can have both positive and negative sides (Tong 1996, 76). Rosemarie Tong distinguishes four main streams of feminist ethics, namely feminine, maternal, political, and lesbian. The unity paradigm of feminist thought is further challenged by postmodern feminism, which questions the very idea of unity and universality of concepts and categories fundamental to feminism, including the category of woman (Tong 1998, 262–264).[6] This useful classification proposed by Tong can be complemented by another view of feminism, in which it is not so much the classification that matters, but the exposure of radical activism and a radical, total commitment to fighting all forms of discrimination and exploitation (Olufemi 2020).

In adopting an eclectic and pluralistic position within the feminist perspective in this book, I reject liberal feminism. The central idea of liberal feminism, the guarantee of equal rights and equal opportunities, is not only impossible under contemporary neoliberal capitalism but also exacerbates existing inequalities, oppression, and exploitation. Capitalism is responsible for inequality as well as environmental degradation. Accepting in the wake of liberal feminism different opportunities for men and women would only mean, as Cinzia Arruzza, Tithi Bhattacharya, and Nancy Fraser note, that not only men but also women can equally participate in exploitation and oppression (Arruzza et al. 2019, 2–3). Liberal feminism does not lead to a change in the structures of exploitation, as it focuses, in line with the idea of capitalism, on individual responsibility instead of an analysis of the structure independent of the individual.[7] The structure shaped by capitalism is preserved but is slightly modified and diversified by including a small number of representatives of traditionally marginalized groups, especially women (Arruzza et al. 2019, 11).[8] The space environment has included women in missions but is still dominated by men and the male point of view, as well as ergonomics and many other solutions. Liberal feminism with its demand for equal rights will come of its own accord in an ideal society. However, in the current non-ideal one, what is important is to change the structure, eliminating sexist and patriarchal inequalities and injustices. If girls are raised differently from boys even in Western societies that have experienced feminism as a significant cultural period, then equal rights for women are not enough for such traditionally raised women to be able to realize their potential.

Feminist bioethics can be practiced in at least two ways. Historically, feminist bioethics has focused on issues considered typically feminine, or issues that affected women in specific ways, as well as impacting women more than men. The main focus of feminist bioethics in this sense has been on female reproductive biology, as well as the concept of woman as mother and the category of care attributed to women or associated with women more than with men (the ethics of care). This is the former way of doing feminist bioethics, when feminist bioethics was understood as focusing on women's issues, that is, in practice, almost exclusively on reproductive biology. It is now "known" that feminist bioethics must go beyond those frameworks that were initially associated with feminist bioethics and women, and apply to essentially all areas of bioethics (Dodds 2004, 2). It is this second way of practicing and understanding feminist bioethics that means applying categories such as sex, gender, domination, exclusion, power, and others fundamental to the feminist perspective.

In a sense, the parallel between the development of feminist philosophy and feminist bioethics is justified. Feminist philosophy was originally focused not only on the issue of women's rights exclusively, but also on the issue of white women, mostly middle class. Black feminists were critical of mainstream feminism, which equated the oppression of women in general with the oppression of white middle-class women who were tired and bored with domestic life (hooks 2001, 33–34). Moreover, many black feminists recognize the harmfulness of white feminism, which, in their view, portrays black feminism in a caricatured and harmful way, and which continues to focus exclusively on sexism as the sole source of oppression (Chaddock and Hinderliter 2020; Schuller 2021). Without feminists of color, the progression in morality, understood as the expansion of the moral circle to include previously excluded groups, that nonfeminist philosophers such as Allen Buchanan and Rachel Powell (2018) are talking about today, could not have happened.

Later, there was an interest in women of color as well as other excluded groups. The category of women was replaced by the category of the excluded, and the limits of oppression and marginalization extended to everyone, not just (white) women.[9] This progression from an exclusivist to a more inclusive approach can also be seen in feminist bioethics. An illustration of this progression is this book, which applies feminist bioethics not only to all marginalized and excluded groups, but also to a new area not previously discussed within feminist ethics and bioethics, namely space exploration. Using feminist philosophy and bioethics in this book, I assume that various social, political, and bioethical actions can lead to negative consequences and oppression of many people, sometimes all of humanity, or at least groups other than just women. Nevertheless, I also accept, following Susan Sherwin, that gender bias is still prevalent and therefore it is useful, when analyzing a chosen ethical situation, to begin the exploration by identifying possible gender biases (Sherwin 1992, 56). Similarly, the concepts of bioethics are being gendered (Rogers et al. 2022, 2), and one of the goals of this book is to highlight how and where this gendering of concepts is taking place in bioethics in relation to future space missions.

Feminist approaches to space exploration in general, and to the bioethics of space missions in particular, imply several issues. First, they draw attention to the role of women in this endeavor as almost absent from space as subjects of action. Second, they signify a feminine perspective and a female, rather than male, viewpoint on what space exploration, including the biomedical issues associated with that exploration, should look like. Third, a feminist approach means paying attention to other marginalized and unaccounted-for groups. Their discrimination runs around categories such as race,[10] ethnicity, class, but also ableism.

Therefore, this book examines the issue of feminist bioethics of space missions in relation to the following two areas. The first area is an analysis of the broader context that sometimes goes beyond bioethics. This context is organized vis-à-vis key categories for feminist critique such as power, oppression, discrimination, subordination, and exclusion. A second context for the analysis is to focus on purely bioethical problems in space and how feminist approaches can be useful in addressing them. In this book, I discuss issues such as disability in space and the concept of human enhancement, as well as the issue of antinatalism in relation to the broader global and environmental context.

I treat feminist philosophy in this book as equivalent to normative ethics. It is therefore not merely an alternative approach, not a contextual ethics, as the feminist perspective is often mentioned alongside the narrative method or casuistry. Feminist ethics and bioethics are no less important than duty-based ethics and utility-based ethics. Moreover, feminism is a grand, all-encompassing philosophical system that proposes, in our opinion, an optimal approach to the problems of both earthly reality and, most importantly, space exploration. Feminism understood as a philosophical concept and at the same time as an ethical and bioethical system has many advantages. Feminism as a concept means the abstract collection of all our intuitions and perceptions about feminism, which, despite their differences, find common ground in concern for the excluded and the fight against social injustice. Feminism as a system, on the other hand, signifies an elaborate, nuanced theory, containing the often differing perspectives of different feminists on important issues, in the case of interest to us, bioethical and biomedical issues. However, it seems that the most important of these is to point out that the way we think about ethics, as well as our ideas about how moral reasoning takes place and should take place, are closely tied to social practices and political phenomena (Mosko 2018, 1366).

The Epistemic Significance of the Particularistic Standpoint: The Gender Controversy

Feminist philosophy and bioethics emphasize the importance of subjectivity. The category of standpoint in feminist epistemology points to the role played by the particularistic point of view. Standpoint epistemology indicates that cognitive subjective factors, which are treated as non-epistemic in traditional nonfeminist epistemology, have epistemic value (Toole 2021). It is indicated here that there is no single objective point of view, but the point of view is different for women and for men. To this should be added specific viewpoints for other groups, for example, sexually nonbinary, or privileged or under-privileged by race, ethnicity, or class. This problem can also be presented as a problem related to considering hitherto male philosophy as the only philosophy and granting it the status of universal. Such a philosophy, however, while it may express some implicitly universal human experience, is not the only universal human philosophy (Holland 1990, 2). Western philosophy dominated by the so-called male point of view is one of the causes of epistemic oppression regarding the perspective of non-dominant groups both in philosophy and everywhere else outside of it (Toole 2019).

But the differentiation is not only along gender lines, for it also applies to other social categories. Moreover, there are differences within a group. Radical feminism emphasizes the epistemic privileging of women due to their status as victims of a repressive system. Cultural feminism, on the other hand, exposes another aspect of women's epistemic uniqueness, namely their attachment to relationality, connectedness, and a sense of caring, which, according to cultural feminists, gives women a unique epistemic perspective in those relationships that involve the category of caring. As Tong notes, both approaches are simplistic visions that fail to take into account many nuances and intersectionalities. These perspectives are worth supplementing with a postmodern epistemic critique that denies the existence of any unified, universal standpoints in favor of a singular standpoint unique to each individual (Tong 1997, 88–89).

A specifically female experience—even if not shared by all women, but one that is characteristic only, or primarily, of women—is the experience of submission. Being submitted in the sense analyzed by Manon Garcia means voluntarily submitting to male domination. Voluntary submission to someone contradicts needs to maintain freedom and autonomy. But as Garcia argues, an oppressive male-dominated society, with the aid of cultural texts rooted in religion and philosophy, has cultivated and continues to cultivate in many places today an image of women as beings either inherently submitted to men or destined to become so (submission as a kind of feminine moral virtue) (Garcia 2021).

Garcia, expanding on Simone de Beauvoir's philosophy, emphasizes the usefulness of Beauvoir's category of "situation." The "situation" of women means that for various social, political, and cultural reasons, a woman's situation is different from that of a man. This is not due to biological differences, but specifically to the social position of women (Garcia 2021, 54). In practice, a woman's situation means being in a worse position at the starting point than the analogous position of a man. This positioning of a woman affects her possibilities of choice and action in the area of freedom because it imposes limitations and impediments that cannot be so easily surpassed in the name of individual absolute freedom, which Jean-Paul Sartre ascribed to the individual. Consequently, even if we reject biological determinism and essentialism and accept that sex differences cannot account for differences in opportunities and social positioning of the two genders, we live in a reality in which sex differences are essentialized and determine social positioning and opportunities in ways specific to women and to men. Sex, or biological differences such as hormones, chromosomes, and sexual organs that determine female or male categories, became the starting point for the oppressive application of gender categories to women, prescribing or suggesting certain patterns of behavior. A woman must be "woman" because of her biological sex. Therefore, one of the feminist ideals has become a society composed of different sexes, but genderless (Mikkola 2016, 3, 22–24).[11] Although biology does not provide a basis for deriving gender differences (Beauvoir 2011, 47), sexism is often rooted in biological differences, which are elevated to the status of forces determining the fate of men and women (Savigny 2020).

But gender is not the only social product. Sex, too, is in many cases socially constructed. Despite the fact that some people cannot be clearly classified as male or female, there are usually no other categories to describe individuals who cannot really be classified into one of these sexes. Another example is defining sex based on arbitrarily chosen criteria, as in sports (Kramer and Beutel 2015, 3–4). The philosophical understanding of gender and sex should be constantly supplemented by the perspective of gender sociology, which emphasizes the multiplicity of expressions and realizations of masculinity and femininity, which can take on different expressions according to social position and role (Kramer and Beutel 2015, 14). The categories of sex and gender are not binary, but they are gradable. The category of sex, for example, straddles one extreme, which can be considered to represent so-called typical male and female traits, and another extreme, which is a variety of traits that cannot be classified as either male or female (Hendl and Browne 2022, 157).

The traditional nonfeminist association of gender roles with sex characteristics and functions applies not only to behavior and clothing style, but more importantly to the division of labor. The division of labor is typically gendered in many communities, and it is this division that determines the social and private roles of men and women (Dupré 2017, 228–229). However, as John Dupré points out, biology does not provide a basis for essentialist thinking that assumes the fixity of categories such as sex. This is so for at least two reasons. First, as Dupré points out, vast numbers of living organisms have no sex at all. Many other animal species change sex during their lives. Others in turn have more than two sexes (Dupré 2017, 231–232). Second, life is a process, not a thing. Dupré and Daniel J. Nicholson reject a static understanding of living organisms in terms of essences and things in favor of a processual understanding of all living organisms in terms of constant change. What is essential are life cycles, not properties (Dupré and Nicholson 2018, 11–20).

Using the processual philosophy of biology, we can say that the essentialism underlying the tendency to derive fixed gender from what is also imagined as fixed sex implies bottom-up thinking, where an arbitrarily and randomly

selected trait is taken to be fundamental and determinative of a larger structure. In the case of antifeminist understandings of sex and gender, such an
analogy to bottom-up thinking is the recognition of selected traits such as
chromosomes or hormones as determining and causative of the entire organism. As Dupré and Nicholson note, the reductionism inherent in thinking
in terms of bottom-up causality occurs in conjunction with essentialism
and mechanicism (2018, 27–28). Arbitrarily designated elements such as
sex organs (reductionism) determine not only sex but also gender (essentialism and bottom-up thinking), and the woman is a machine engaged in
childbearing and domesticity (mechanicism). Criticism of the concept of
biological gender essentialism has a long tradition in feminist philosophy
and emphasizes not only the dominant role played by socialization, but also
the lack of a homogeneous class of woman and the presence of numerous
differences between women (Mikkola 2017).

Not only is sex processual from a biological perspective, but it is also socially constructed, as is gender. According to some feminist interpretations,
societies have certain ideas and expectations about gender. These expectations then influence their decisions and actions regarding sex. This is especially true for infants, where it is not always possible to clearly define their
biological sex. But because societies have only two genders, they tend to strive
to categorize each body in such a biological way that it corresponds to either a
female or male gender (Stone 2007, 46–48).

Despite the impossibility of describing the situation of women in universal terms, it is possible, following Josephine Donovan, to point to certain
common features in the experiences of women from different groups and
cultures. These features account for the generic similarity of women's experience and situation and at the same time for their distinctiveness from men
(Table 1.1).

The division between gender and sex and its essentialization has long-term
consequences for women. One of them relates to physique, strength, and fitness. Iris Marion Young bases her considerations on Beauvoir's category of
situation as well as Maurice Merleau-Ponty's concept of the lived body. Young
points out that it is indeed possible to observe differences in body motor skills
between women and men (girls and boys). These differences are only seemingly a result of biology. Drawing on the aforementioned French philosophers,
Young notes that the social environment alone threw women into such situations that it forced and expected them to operate and use their bodies in a
certain way (Young 2005, 31). Young further elaborates that women's situation, unlike men's, consists of their oppression, which results in the existence of modalities specific to women regarding their use and experience of

Table 1.1 The experiences of oppression and exclusion common to women of different cultures and social groups

Women	Men
Political oppression: lack of real influence over political decisions and events. Marginalization and less serious treatment of women in leadership positions than men in equivalent positions.	Real political influence
Housework: women are traditionally more likely than men to be involved in childrearing and household duties.	Men spend incomparably less time on housework. It is still associated with the domain of women, which is evident in the way girls are raised differently than boys.
Subsistence economy: women produce incomparably more often than men not for sale or exchange (the abstract value of the labor product), but for direct consumption.	The work done by men used to be treated as more important than that of women, which is often evident even today. In contrast, traditionally women's domestic and care work is unlikely to be treated as work. Women's work is associated with more monotony and repetition, while men's work is associated with originality and creativity.
Physiological processes: all or the vast majority of women experience processes unique only to women, such as menstruation, pregnancy, breastfeeding, menopause, and other specific experiences of adolescence.	Men do not experience any physiological processes typical only of men, which would have the consequence of monitoring, policing, controlling, or stigmatizing them.
Male violence: exposure to rape and physical violence by men	Men are often victims of violence committed by other men, but they are free from the permanent risk, in particular, of sexual and domestic violence, which are typically experienced by women.

Source: Donovan 2012, 168–169.

their bodies. Consequently, traditional feminine phenomenology of the body includes such elements as a belief in, and sense of, one's fragility and delicacy, a tendency toward immobility rather than mobility, and a fear of injuring oneself and getting dirty. All of these environmentally conditioned components lead to a phenomenologically female-specific operationalization of their bodies that, compared to a male approach, has a character of insecurity, withdrawal, and closure (Young 2005, 42–44). It can be said that the categories of prisoner, and partly of disabled person, illustrate this socially imposed self-limitation of one's bodily and motor capacities.

As many feminists argue, women have always lacked their own identity as subjects (Anderson et al. 2021). The subject has always been the man; the woman has functioned only in relation to the man with the status of an exemplary human being, or simply identity with the man ("man" as human being and "man" as man [male human being]). Also, traits associated with women and femininity, including women themselves, have been evaluated from a male hierarchical perspective as lower than the so-called masculine traits of character and intellect (Stone 2019, 26). Pairs of opposites fundamental to European culture and philosophy such as culture-nature, mind-body, and reason-emotion recognize the primacy of the first term and identify it with the domain of men, while the opposing terms, as the domain of women, are rated much lower.[12] The consequences of this dichotomy are political and involve the exclusion of women from the public sphere. Moreover, one can say, following Luce Irigaray, that Western culture is matricidal because it relegates everything related to motherhood to the private sphere. Consequently, leaving the sphere of motherhood has become a condition of participating in civilization (Stone 2019, 26–27).[13]

Women in the Light of Biology and Evolution

The problem with understanding and analyzing women as a separate category is evident in the philosophy of science and science itself (Tomm and Hamilton 1988).[14] Feminist philosophy of science draws attention to the problem of stereotyping women, as well as the application of gendered concepts and theories.[15] This is also evident in the evolutionary sciences, including psychology and evolutionary biology. Feminism, with some exceptions, is generally skeptical of the conclusions of the evolutionary sciences, especially sociobiology and evolutionary psychology, insofar as these disciplines assume that certain behaviors and modules that are different for men and women are biological adaptations.[16] One example of such theories is "staying alive" theory (SAT), which assumes that women have biological adaptations that make them more cautious than men in various areas of life such as fighting and aggression, but also attitudes to germs, hygiene, and illness that are linked to their function as mothers and caregivers (Benenson et al. 2022). SAT has the potential to reinforce stereotypes. It preserves old-fashioned notions about women and ties them to their supposed subordination to raising children. If we take the perspective of a feminist philosophy that assumes that certain psychological and behavioral traits that we more often attribute to women have social rather than biological roots (Mikkola 2017), SAT then becomes an example of essentialist thinking about women. After all, it may be that women exhibit certain behaviors more frequently solely because they have been forced to do so as a result of centuries of oppression and domination by men (Lindemann 2019).

Thus, an unjust social arrangement has generated the regularities described by SAT. SAT excludes this possibility by equating the woman with the mother at the starting point.

This theory excludes not only the dynamics of homosexual couples but all persons other than men and women. SAT reinforces the stereotypical role of women as caretaker mothers, emphasizing the importance of a woman's survival rate for raising offspring. This supposed biological regularity is the reason for the exploitation of women not only in the domestic and family circle, but as caregivers in various areas of social life. Gender is reinforced and constituted through the repetition of gender-coded acts. As Judith Butler points out, gender is a kind of performance where we are taught from childhood to play the roles assigned to us, either female or male (1990).[17]

Essentialism is also evident in the philosophy of biology adopted by SAT, which seems to assume that sex differences between men and women are fundamental, fixed differences that determine the described patterns of behavior that are different for both sexes (Weaver and Fehr 2017). Alison Stone points to the dominance of the two-sex model in philosophy of biology and philosophy of science. This model exposes the differences between men and women and focuses less on finding and revealing the similarities between them (Stone 2007, 36–37).

It is also worth noting another SAT conclusion that is paradoxical. Women have greater chances of survival and a longer life expectancy, but a lower quality of life due to more often experiencing fear and pain, as well as more frequently finding themselves in social roles that cause discomfort. This fact conflicts with a philosophical tradition that emphasizes the importance of well-being, the so-called life worth living. SAT shows that women inherently have lives less worth living than men, a paradoxical adaptation as long as we value quality of life rather than existence itself. This seemingly more advantageous situation for men as measured by a higher degree of well-being can be compared to Derek Parfit's (1984) famous thought experiment about hypothetical future humans who live a life worth living for forty years, then die as a result of radioactive waste stored by an earlier generation. But before they die, they do quite well.

It can be assumed that women's lower quality of life results from centuries of oppression and domination by men, and thus from a pathological social system. From this point of view, longer survival becomes a dubious advantage, further reinforcing the dominant position of men, who are freed from the possible "duty" of caring for offspring by longer-living women who live longer to care for their offspring. This reinforces the feminist diagnosis of social relations in which men, even if they live shorter lives, enjoy life more than women. Joyce F. Benenson and colleagues (2022) do not explain causation, but references to similar regularities in other species suggest that they recognize a biological, rather than environmental, determination of regularities such as greater concern for offspring in females than in males.

The essentialist thinking responsible for the persistence of sexism is present in both colloquial thinking and scientific discourse, as well as philosophy and bioethics. Essentialism assumes the existence of essential and universal qualities that are shared by all women. Consequently, women can be understood as one and the same kind that share the same natural and/or cultural characteristics (Stone 2004, 86–87). Strategic essentialism states that we should accept the idea of essentialism for the sake of political pragmatism even if descriptively it is false. The concept of the notion of women as a series, on the other hand, assumes that while ontologically there is no common characteristic that unites women, there is a commonality of a series of events and regularities, a common structure that accounts for the same or a similar experience of women (Stone 2004, 88–90).

Care Ethics

Women have traditionally been associated in sexist and patriarchal societies with caring for others, which can also be seen in the scientific theories presented above. Feminism, too, has not remained indifferent to this phenomenon. The ethic of care has an important place in feminist bioethics, although it should not be equated with feminist bioethics. Feminist bioethics draws attention to the importance of caring but stresses the risk of exploitation and essentialist approaches to gender if greater caring is equated with femininity.[18] Feminism also draws attention to the risk of seeing the relationship between mother and child as a model of caring. An ethic of care is the foundation of human life, of humanity as a whole, for the upbringing of children, as well as the care of all those in a relationship of dependence, requires precisely the care that in practice is usually taken care of by women, and this care is undervalued in mainstream bioethics. As Alisa L. Carse and Hilde Lindemann Nelson argue, the well-being and development of society are based on the virtues inherent in the ethic of caring, associated with femininity, namely benevolence, compassion, and kindness, as well as the ability to cooperate and mitigate conflict (Carse and Nelson 1999). Despite the strategic importance of the daily care provided by women around the world, which the capitalist system is well aware of (Cameron 2020), this care has a low social status and is rarely remunerated at all, or if it is, it is usually for the minimum possible wage.

Thus, we can say that the ethics of care is, on the one hand, one of the key propositions of feminism in ethics and, on the other, a feminist reaction to the abstract masculine ethics based on principles and rules. But at the same time, the ethic of care carries the risk of deepening stereotyping, although this risk arises not so much from the very nature of the ethic of care but from its application in an oppressive, discriminatory society for women. One of the proposals of the feminist ethic of care is to construct a concept of relationship that refers to the relationship of friendship as a model, instead of that between the mothering person and the child, which is a relationship between unequals (Tong 1997, 48).

Care ethics is important for another reason. It expresses what is central to feminism, namely the belief that there are differences between men and women; however, this is a paradox within feminism leading to a dispute between the feminism of differences and the feminism of sameness. Forerunners of ethics of care such as Carol Gilligan and Nel Noddings emphasize that the default mode of moral reasoning appropriate to women is one of relationship, connection, and care, in contrast to the male default approach to moral reasoning, which is abstract in nature and operates on general principles and rules, with the principle of justice at the forefront (Tong 2009, 165–168). The very idea of linking the concept of care not so much to feminism as to femininity grows out of the life experience of women who, due to patriarchal social structures, had and often still have today the role of caring for children and the elderly, and often of caring for and supporting men. Thus, the life experience of the statistical woman is not that of an autonomous, independent, relationship-separated individual. A woman's experience is usually one of responsibility for others and caring, so it is difficult to expect women to consider an ethic based on the concept of independent individuals to be appropriate and "their" ethic (Sherwin 1992, 47). Another issue is the moral evaluation of the concept of caring associated with women and femininity. Some feminists believe that it is worthwhile exposing this component, while others see it as an example of exploitation of women and reinforcement of the traditional role of women as a service to the rest of society.

If we accept the existence of this distinction in moral reasoning, it will lead to differences in moral conclusions and ethical evaluation of the same situation, depending on whether the person evaluating the situation is a woman or a man. In a society dominated by a masculine style of moral reasoning, the consequence of this distinction is to marginalize the female perspective. But, perhaps paradoxically, under the conditions of an ideal society in which both men and women effectively have equal rights, opportunities, and influence, the problem of balancing two such different modes of moral thinking will remain. The ethics of care illustrates the dissimilarity of two perspectives in looking at the situation of women that characterizes liberal feminism and radical cultural feminism. For the former, what is essential is the guarantee of equal rights and opportunities. This is a necessary starting point, but under the conditions of a capitalist society (and even a socialist society, as socialist feminism points out),[19] women are more often than not in a worse situation than men, which contradicts the thesis of the sufficiency of conditions of consent and choice. If we take a liberal view, will the difference in moral reasoning between men and women, as indicated by the ethic of care, under conditions of actual equality of rights, chances, and opportunities, actually guarantee women the same opportunities and chances that men have with their abstract view of morality? Probably not.[20] Therefore, perhaps a rational view is the proposal of radical cultural feminism, which exposes the ontological differences between men and women and seeks to use the characteristics inherent in women—derived from female reproductive biology—as women's assets.

Ethics of care proposes a new view of the functions of ethics. The task of ethics is not to establish criteria for correct action or to reflect on the moral nature of human beings, but rather to help people who are suffering physically and psychologically. This task requires an attitude of relationality and connection (Tong 2009, 173).[21] The ethic of care also postulates a specific agenda for action in public life. In contrast to the masculine patterns of thought and action based on competition; selfishness; violence and aggression, to a degree; and individualism, a feminist ethic of care promotes attention to emotion, sympathy, and empathy, in a global and environmental context (Tong 2009, 195–199). However, regardless of the specific moral qualities attributed to women in the tradition of care ethics, Gilligan points to the need to see, understand, and accept the female voice and female perspective, which is often not only overlooked, but even misunderstood and unconscious in a patriarchal society (Gilligan 1997).

The search for a feminine ethical element to be the domain of women alone runs the risk of reinforcing patriarchal structures. Sarah Lucia Hoagland argues that the popular image of a woman is that of a mother, even if it is not to take care of her own children but to be responsible to the world. However, in both patriarchal societies and those in which old patriarchal ideas still remain strong, the concept of motherhood is often linked to ideas of servitude, subordination, and self-sacrifice (Hoagland 1990, 285–289). In future long-term space missions, the inclusion of women may be motivated by this linking of women to their function in patriarchal societies of caring and nurturing. In the confined and isolated conditions of space missions, it is particularly important that the preparation and selection of candidates also take into account the predisposition to care for others equally possessed by, and required of, all genders. This is certainly a social model that participants in such missions could wish for. But it is impossible to plan such a social cross section of missions without engaging sexist biases in thinking of women as performing just such socially serviceable functions. Adequate training, as well as selection for an expected psychological and behavioral profile independent of gender, should replace sexist thinking about characteristics attributed to women but socially attractive. The selection of candidates for future long-term space exploration, devoid of sexist biases, should be preceded by a modification of social relations on Earth, where, starting from the upbringing of children, girls will not be molded into social caregiving roles. Only by eliminating this shift of the element of caring and nurturing from girls and women equally to boys and men would it be possible to eliminate the risk of such bias during selection for future space missions.

Ethics and Bioethics Centered on the Category of Power

The opposite of the ethic of care associated with women outlined above is the ethic of power considered to be the domain of men. Since the category of power is one of the most important categories in feminism, the issue of men's power over women and the unjust distribution of power in general, not only in society but also in various theoretical concepts, is relevant to feminist ethics (French 1992). In ethics centered around the category of power, the starting point for ethical analyses is the establishment and identification of men's power relations over women, structures of oppression, domination, and subordination (Tong 1997, 48–49). Feminist bioethicists point to the structural exclusion of groups such as the sick, the "other" (that is, other than the dominant group) (Schulz and Mullings 2006), medical personnel other than physicians, and women (Holmes 1999).

Susan Sherwin's vision of feminist bioethics emphasizes that the latter should analyze power structures in healthcare as well as discourse about it. Sherwin argues that the dominant analytical perspective in nonfeminist bioethics is that of the physician to the exclusion of both the perspective of the rest of the personnel involved in medical care and that of the patient. This suggests a recognition of the dominant role of the medical profession currently reserved for men.[22] Moreover, Sherwin notes that even when the other marginalized groups are discussed, bioethicists' attention is focused on concepts and issues related to power and control, such as autonomy, paternalism, and the informed consent rule. Therefore, the role of feminist bioethics is primarily to deconstruct power structures, interest hierarchies, and male bias (Sherwin 1992, 3–4).

Feminist bioethics offers a new methodological approach, including an alternative model of ethical decision-making. This model assumes situationality and positionality in interpreting moral principles and moral rules. Because there are differences among people, including all parties involved in making ethical decisions about medical and biomedical issues, caused by the degree of power held, which is influenced by gender, race, class, and other factors, individuals may perceive the meaning of particular ethical principles differently. The feminist model of ethical decision-making combines a rational perspective with an emotional-intuitive one, accounts for differences in power structures, and takes into account the presence of cultural biases (Hill et al. 1995, 21, 27).[23]

Individualism and Abstractionism in Nonfeminist Bioethics

Nonfeminist bioethics in its classical paradigm is usually principle based. Principlism is criticized not only by feminism, but also by many other nonfeminist bioethicists who insist that bioethics must be issue driven, not principle driven. In any case, it is primarily the feminist critique that emphasizes the inappropriateness to real-world situations of bioethics assuming the existence of abstract principles and then interpreting each ethical situation through the prism of those principles, usually through a conflict between principles (Holmes 1999).

But the feminist critique of nonfeminist bioethics is not just about methodological issues. The objection is much deeper, for it concerns how basic ontological and epistemological concepts, primary to the methodologies employed, are understood. Feminist bioethics criticizes the ontology of nonfeminist bioethics, which is based on the concept of an individual who is an abstract being, in a sense understood as if deprived of a body and devoid of social relations. The correct understanding of the individual that reflects real circumstances is an understanding that exposes corporeality, emotionality, and dependence, as well as relationships with other people (Scully 2021, 274–275).

Feminists often reject the understanding of the body and the individual inherent in liberal individualism, in which the body is not part of identity, but is possessed by the mind. This transcendental understanding of the body not only ignores the realities of embodiment but also represents a reality historically and socially closer to men than to women. It was men, as Elizabeth Frazer and Nicola Lacey note, who had incomparably greater opportunities to live a life detached from the many responsibilities seen as the domain of corporeality and could concentrate on the intellectual and spiritual realms (Frazer and Lacey 1993, 53–54). Such an opportunity may indeed have given the impression of separating the body from the individual, identified with the mind. Female reproductive biology has often limited women's opportunities and interrupted or prevented their professional development—something men have not experienced because of their biology.

Thus, nonfeminist ethics and bioethics are unrealistically focused on considering judgments and choices, whereas the full moral picture should additionally also include interpretations, descriptions, communications, narratives, and other activities that are part of the human web of relationships and commitments (Scully 2008, 46). What does this feminist shift of focus from the abstract, individual, and rational subject to the concrete situation and circumstance mean in practice, and what does it entail? It means that feminist bioethics draws attention to possible differences in moral judgment and moral reasoning in different individuals, due to their status, life situation, and a host of other factors. This particularistic approach stems from the assumption that there are differences between people—moral subjects—in their moral understanding of ethical situations (Scully 2008, 47). In contrast to this feminist perspective, nonfeminist bioethics assumes universalism.[24]

Feminist critique of bioethics thus rejects abstractionism and individualism, emphasizing that we are always dealing with a concrete human being who lives in concrete circumstances and a concrete system of connections and dependencies (Marway and Widdows 2015). Therefore, feminism criticizes all major normative systems such as ethics based on duty, consequences, or contract for their reliance on what is universal in all human beings, which is only an idealized conception of the universal human being, assuming her autonomy, rationality, and self-interest. These systems also marginalize the value of what is believed to be the domain and reality of women, namely the realm of personal relationships and individual responsibility (Sherwin 1992, 38–42).

The feminist constructive response to the critique of individualism, abstractionism, and power structures was to propose the concepts of relational autonomy and self, particularity, and the importance of justice as a counterweight to power (Marway and Widdows 2015).

There is no doubt that the ontology of the autonomous individual possessing autonomy plays a larger role in nonfeminist ethics than in nonfeminist bioethics. Feminism criticizes traditional philosophical and ethical ontology for presenting a dualistic vision of the world. The self is understood in opposition to other individuals. The nonfeminist "self versus others" ontology is replaced by a feminist "self in and through others" ontology. One possible model might be the mother-child relationship, which is an example of the strongest symbiotic relationship in nature that is opposed to ideas of individualism, independence, and autonomy (Tong 1993, 51). As indicated above, this model has its limitations and risks, but it nevertheless reminds us of a basic type of dependency inherent in almost all people at least at some stage in their lives, which nonfeminist philosophy and bioethics, too focused on abstract individualism and autonomy, seem to have missed.

Any general bioethical theory must remain imprecise, and no general theory can offer detailed solutions that explain every bioethical situation unambiguously and avoid conflicting principles and rules (Beauchamp and Childress 2013, 395). The four famous bioethical principles are at a high level of generality and require specification each time. Feminist bioethics, in a sense, is a normative system that has emerged from the beginning as a system dynamically subject to specification each time. Feminist demands for the protection of the excluded and of care are in a sense a specification of principles of justice and beneficence. The particularism inherent in feminist bioethics is possible only on the basis of a bottom-up methodological approach, in contrast to the top-down approach characteristic, for example, of principlism and other normative systems based on duty or the principle of utility. Feminism applies a bottom-up methodology. From this point of view, it may be said that the feminist critique of abstractionism inherent in nonfeminist bioethical systems concerns not so much the systems themselves as the methodology adopted.[25]

At the heart of feminist bioethics is the belief that moral principles and rules should never be considered in the abstract, without taking into account the context of their application (Sherwin 1992, 77). This context includes their genesis as well as the power structures, that is, the parties who gain and lose from the application of the principles and rules in question.

A Feminist Critique of Nonfeminist Ethical Normative Theories

In light of the outlined difference between feminist and nonfeminist approaches to bioethics, it can be seen that one of feminism's main objections to nonfeminist normative philosophy and ethics is the charge of abstractionism.

Normative theories are detached from reality; they do not describe people or the relationships between them as they actually occur. Paradoxically, this abstractionism, in a sense, is not entirely abstract. As the feminist critique of normative systems shows, the proposed theories express the point of view of a particular group. That group is privileged men. The thinking inherent in systems such as deontology or utilitarianism to some extent, despite its abstractness, expresses a male standpoint. It assumes that autonomous individuals are the subjects of moral decisions. The male standpoint understood in this way has a very exclusive and elitist character. This problem is visible, among other things, in the concept of prima facie and actual duties proposed by W. D. Ross. As Tong points out, the list of prima facie duties proposed by Ross may indeed be intuitive, but they are intuitions shared within his social circle. The moral intuitions of a particular social circle do not necessarily express the intuitions of the rest of society. This leads to the risk of normalizing one point of view and accepting it as the norm, at the cost of devaluing the points of view of other social groups (Tong 1997, 18). A similar objection can be made to the concept of common morality of Tom L. Beauchamp and James F. Childress, which can also be the common morality of a privileged social class in Western countries.

The critique of abstractionism in ethics and bioethics also extends to the critique of abstractionism in social and political philosophy. An example is the legitimate practice of defending human rights as women's rights and with women in mind, but which, as long as it is focused on an individualistic and abstract understanding of human rights, such as property rights, and does not take into account the specific social and historical context, can paradoxically reinforce the inequality and exploitation of women (McLaren 2019, 96). Margaret A. McLaren proposes complementing the rights-based model and individual choice with a feminist social justice model that goes beyond abstract rights and also takes into account existing inequalities, oppression, and power structure dynamics (2019, 97).

In summary, nonfeminist normative ethics is abstract, does not take into account the actual complexity of ethical situations, is not objective, tends to represent privileged groups, and contains gendered concepts.

A Critique of Classical Bioethical Principles

Feminism criticizes nonfeminist bioethics primarily for its domination by principles and rules. *Principles of Biomedical Ethics* by Beauchamp and Childress (2013) (the first edition in 1979 has seen eight editions to date) is traditionally cited in almost every feminist critique of the methodology of nonfeminist bioethics. Susan M. Wolf critiques the following four core elements of nonfeminist bioethics: the aforementioned dominance of principles and rules at the expense of recognizing the role played by diversity and context; the validity of liberal individualism and the failure to recognize the importance of the group; the subordination of bioethics to hierarchy and institutions at the expense of the excluded; and the methodological isolation of bioethics from dominant currents in philosophy such as feminism and postmodernism (Wolf 1996, 14). The four main principles in bioethics, namely respect for autonomy, justice, beneficence, and nonmaleficence, are criticized from a feminist perspective for, among other things, not setting boundaries for the application of the category of medicalization, not taking into account the patient's standpoint on medicine, and not considering the specific context of the needs and problems of individual patients (Wolf 1996, 21).

Nonfeminist mainstream bioethics incorporates some of the issues discussed by feminist bioethics and, more broadly, feminist ethics. But this does not yet make nonfeminist mainstream bioethics a feminist bioethics. For example, the fact that the aforementioned Beauchamp and Childress addressed feminist criticism and pointed out the problem of contextuality does not make them, like many other mainstream bioethicists who point out, or rather signal, the existence of contextuality and particularity of situations, feminist bioethicists. For this to happen, bioethics must, as Anne Donchin and Jackie Scully contend, be focused on gender and social justice categories (Donchin and Scully 2015). In other words, these categories should be the central categories for feminist bioethics, as opposed to the aforementioned classical four principles and any other normative theories. These categories are the main conceptual tools of feminist bioethics, which asks questions about who gains and who loses in a given bioethical situation.

Bioethics based on the principles mentioned above is primarily too abstract. For example, the principle of respect for autonomy does not take into account the entanglement and dependence of the individual. But this abstractness of the principles, which does not reflect the real position of many individuals, is at the same time gender biased. Ostensibly, abstractionism and gender bias are two separate issues, but in practice they remain related. For both represent the viewpoint and position of the white middle-class male (Sherwin 1996, 53–54). As feminist criticism argues, this abstract individual who has autonomy and who can afford to take independent and just decisions and actions is precisely the independent white male.

Principles in bioethics can function in an authoritarian manner and dominate discourse. The consequence of their dominance may be the nonacceptance of alternative ethical judgments. Simona Giordano even compares some contexts for applying principles to the authority of religion, particularly in past debates about morality (Giordano 2010, 47).

In summary, bioethical discussion since the 1970s has been dominated by abstract topics usually discussed from a male (doctor's) point of view, predominantly through the prism of abstract concepts (Bystranowski et al. 2022).

Autonomy versus Relational Autonomy and Relational Self

Perhaps the most important bioethical principle is the principle of respect for autonomy. A basic nonfeminist understanding of autonomy presupposes the following two conditions. An individual can be considered autonomous when she possesses liberty (i.e., independence from external factors) as well as when she possesses agency, that is, the capacity for intentional action (Beauchamp and Childress 2013, 102). Feminist ethics and bioethics offer their own understanding of the principle of autonomy, so-called relational autonomy. Feminist philosophers point out that while autonomy is a value in itself, we can rarely speak of a state of pure, abstract autonomy, which is precisely in this abstract form that is usually presented in nonfeminist philosophy and bioethics.[26] The problem with autonomy identified by feminism can also be presented in terms of a dyad between those who enjoy full autonomy and those who do not for a variety of reasons. Feminist bioethics adds that many situations identified by a nonfeminist understanding of the principle of respect for autonomy in light of the feminist concept of relational autonomy qualify as a lack of autonomy (Stoljar and Mackenzie 2022). The distinction between capacity and status is helpful here in distinguishing the theoretical principle of autonomy (having the capacity to theoretically be an autonomous subject) from the possibility of actually being an autonomous individual in the here and now. Autonomy as status means that, despite theoretical capacity, in practice a person does not have the environmental conditions to realize her autonomy because the environment does not recognize her as an autonomous person or prevents her from doing so (Mackenzie 2021). Feminists point to the contextuality and relationality of the human being, human dependencies and multiple connections, and reject an abstract understanding of the individual.

The concept of autonomy that is familiar from the history of philosophy and ethics, but also from modern bioethics, is usually a way of understanding and experiencing autonomy characterized by a free and independent man, who is usually the only, or at least the final, instance making a given ethical and bioethical decision. It is not an understanding of autonomy and an experience of autonomy familiar to women. A woman who was/is dependent on her male partner/husband for various reasons, or a woman who is pregnant and then becomes a single mother, cannot experience the so-called absolute or total autonomy that only a man can.[27] Consequently, a woman's experience of autonomy must differ from a man's understanding of autonomy for biological (pregnancy) and socioeconomic reasons (the traditional economic dependence of women on men in societies that discriminate against women and subordinate them to men). The oppression of women has led to a situation where the sense of autonomy can be illusory. This is the so-called concept of content-neutral autonomy, in which a woman's decision to subordinate herself to a man, supposedly made under conditions of real freedom and informed consent, is considered autonomous. Even if theoretically these boundary conditions were maintained, a woman's entrenchment in a given oppressive social context may significantly influence her thinking about male-female relations and lead to an apparently autonomous decision (Friedman and Bolte 2007, 90).

The feminist understanding of autonomy derives from the feminist understanding of the self, the latter of which is partially based on biological human development, in which until the first several years of life an individual depends on others for her survival. In the vast majority of cases, human beings are dependent on others, including in the last years of their life. Relationality, being in a relationship, dependence, being exposed, among other things, by maternal ethics and care ethics, but also the concept of the so-called second person, are the components that are emphasized by the feminist understanding of the self (Witt 2011, 122).[28] A state of dependence and powerlessness particularly characterizes seriously ill patients, whom nonfeminist bioethics nevertheless continues to view in terms of abstract autonomy to the point of promoting the practice of advance directives to respect this absolutely understood autonomy in a situation where the patient will have very limited autonomy, if any at all (Lindemann 2021).

A feminist understanding of autonomy also recognizes as rationales motivating self-determined and self-reflective action those factors that Immanuel Kant's conception of rational action eliminates. A feminist understanding of autonomy also recognizes emotions, desires, and passions as reasons equivalent to reason, provided that they constitute an essential element of a person and motivate her actions. Feminism honors perspectival identity, which encompasses the entirety of an individual's decision-making process and action, and thus also includes an individual's emotions, desires, and relationships and commitments (Friedman 2003, 9–12). Autonomy is considered within this broad context of relationships, connections, and pervasive interconnections—with all that is important to the individual. This accounts for the main difference between feminist and nonfeminist understandings of autonomy. The nonfeminist understanding of autonomy presupposes the existence of an abstract subject of moral action and ideal conditions in which the aforementioned elements that are essential to the feminist understanding of autonomy other than reason are separated from the individual. Thus, her relationships, emotional connections, and desires in a nonfeminist understanding should be eliminated.

An example of this abstract understanding of individual autonomy by nonfeminist bioethics is the way abortion rights are debated. The discussion tends to focus on the embryo and fetus and can lead to the impression that the interests of the pregnant person, not to mention her desires, feelings, and experiences, are not taken into account at all. This way of discussing abortion in terms of "rights talk" expresses a masculine view of the world and at the same time conceals this masculine identification with the fetus, not the woman (Holland 1990, 170). A feminist understanding of autonomy opposes such a model, emphasizing that it presents an unrealistic idea of a human being who always takes into account elements other than reason in her decisions and actions, and usually remains in some kind of correlation with reality and society.

The difference between nonfeminist and feminist understandings of autonomy can also be presented as follows. For the nonfeminist approach, autonomy primarily means the state of the subject at the moment of decision and action. Such a perspective favors an abstract, idealistic understanding of the subject. In contrast to this model, the feminist understanding of autonomy denotes the ability of the individual to consciously reflect on and recognize her needs and commitments and act accordingly (Friedman 2003, 99). This approach is necessarily pragmatic, grounded in the dynamic everyday life of the individual, and describes a way of acting and functioning rather than ideal, a priori conditions for decision-making.

Feminist philosophers contend that understanding autonomy as an ideal, abstract state of an independent, atomistic individual is only one of many possible understandings. Such an understanding is the masculine, classical philosophical view of autonomy, which wrongly equates it with independence and separation from others. One feminist alternative is a dynamic and relational conception of autonomy (Friedman 2003, 84).

Representatives of the classical, nonfeminist understanding of the principle of respect for autonomy, Beauchamp and Childress, commenting on the feminist conception of relational autonomy, admit that they are willing to accept it as long as this conception does not eliminate the three main components of autonomy as they propose it, namely intentionality, understanding, and noncontrol (Beauchamp and Childress 2013, 104, 106).

Relational autonomy emphasizes the individual's context, embodiment, and dependence on social and political circumstances. As Anita M. Superson points out, an individual's particular body shapes the way she is socially perceived and treated by other individuals, in terms such as gender, ethnicity, and being overweight, among others (Superson 2014, 305–306).

Feminist understandings of relational autonomy point to the vulnerability of the person. The individual always has a certain network of social and cultural connections and dependencies that often constrain her. The social conditions that constitute the degree of autonomy of the individual are often unjust, causing oppression and subordination. A feminist understanding of relational autonomy is therefore a nonideal theory of nonideal agents, taking into account their real limitations. The real person, thus understood, is unable to fulfill the criteria of autonomy abstractly understood by nonfeminist bioethics, such as self-authorization, self-determination, and self-governance (Mackenzie 2014, 21–23).

A Critique of Nonfeminist Bioethics

In addition to critiquing categories such as autonomy, rationality, individualism, and abstractionism that are fundamental to nonfeminist philosophy as well as nonfeminist ethics and bioethics, feminism also critiques the framework specific to nonfeminist bioethics through the lens of which it interprets the reality of biomedical problems. Mary C. Rawlinson cites the categories of right, property, and power as such frameworks. Consequently, all bioethical problems are discussed from the point of view of a privileged category of people who function as autonomous, rational individuals based on relations resulting from social contracts (Rawlinson 2016, xviii–xix).[29]

Feminist bioethics can be seen as a critique and correction of nonfeminist bioethics with categories such as gender (considering the consequences for women of particular bioethical decisions) and the ethics of care, as well as paying attention to the contextuality and complexity of decision-making processes, which should not always be the result of an abstract and arbitrary rational decision, but rather a kind of compromise and negotiation that takes into account the interests of all parties (Lindemann 2021). In addition to giving consideration to women, feminist bioethics focuses on all other groups of people who are excluded because of their nonbinary as well as other characteristics.

Conceived in this way, feminist bioethics is both a critique of nonfeminist bioethical concepts and a new theoretical offering that proposes interpreting both familiar and new biomedical challenges in light of categories unfamiliar to nonfeminist bioethics.

Similarities and Inspirations between Feminism and Nonfeminist Ethical Systems

On the one hand, feminism is a new type of normative theory in ethics, while on the other, it uses some principles and rules from other normative systems. Laura Purdy emphasizes the similarities with utilitarianism, among other things, in its emphasis on equal treatment of all parties, and its focus on the consequences of actions, which, according to Purdy, is supposed to exclude the dominant role of prejudices (Purdy 1996, 26). What constitutes the central idea of utilitarianism that accounts for its affinity with feminism is its exposition of the importance of human happiness. As Purdy points out, the only way to maximize happiness is to enable people to satisfy their needs (Purdy 1996, 33).[30]

It is worth noting that the critique of nonfeminist bioethics by feminist bioethics does not necessarily imply a rejection of principles and rules. The position of feminist bioethics within bioethics is aptly described by Tong, who points out that feminist bioethics may act as a mediator between the language of justice and rights proper to nonfeminist bioethics and the language of care, responsibility, and needs proper to feminist bioethics. The two approaches should not be seen as mutually exclusive, but as complementary (Tong 2004, 101).

Thus, to sum up, it can be said that feminist bioethics is concerned with the same principles and rules as nonfeminist bioethics but at the same time emphasizes the constantly nonideal nature of reality and points out that the application of particular principles always takes place within specific, often unjust, power structures.

Notes

1. This book offers a pluralistic approach to feminism, opposing the approach that proclaims the twilight of feminism, and seeks a new approach (Walters 2005). I emphasize the

multiplicity of viewpoints and perspectives within feminist philosophy and bioethics, which must also take into account the non-Western feminist perspective that is troublesome for Western views (with all the risks of being accused of missionary feminism and paternalism; see also Walters 2005), the multitude of different standpoints among women of different races, places of residence, cultures, religions, and their differing perspectives on the same issues, as well as on what should be the focus of feminism. The term "eclectic feminism" (Sherwin 1992, 32) expresses my approach well. Like Emily Maguire, I see feminism as a "multi-functional tool" that allows us to explain any problem with feminism-specific terms and concepts (2019, 8–9).

2. The traditional understanding of patriarchy means the inheritance of honors, titles, and wealth in the male line, and it applies to societies that convey more power to men than to women (Haslanger 2020). Sally Haslanger narrows down the application of the concept of patriarchy to the Western family model, which she says does not describe the situation globally in terms of class and race, where many men are exploited. According to Haslanger, patriarchy is one of the many systems of oppression of which men are also victims, and there is certainly neither one main nor one overarching system of oppression. Haslanger warns against focusing too narrowly on patriarchy and equating it with social evil in general, as such an identification can lead to an underestimation of other systems of oppression (Haslanger 2020).

   Despite some devaluation of the notion of patriarchy in feminist philosophy, in the wake of Robin Dembroff I recognize the usefulness of this notion arising from the fact that the distribution of benefits and privileges often follows gender classifications in many places and situations today, or, in Dembroff's terms, it is more about the domination of the "real man" over everyone else in society. Dembroff thus offers an updated, modernized definition of patriarchy that goes beyond the gender binary-specific approach and points out that many men also fail to meet the real man standard (Dembroff 2024). Dembroff's position resembles the views of Sherry B. Ortner, who says that a real man in patriarchy is a man who realizes the intersectional ideal of manhood, which includes heterosexuality, white supremacy, and able-bodiedness (Ortner 2022).

3. In the book, I present a broad and inclusive understanding of feminism as a movement aimed at achieving liberation and equality for all people (Baumgardner and Richards 2020, 51–52). Some authors stress that we are still dealing with patriarchy, and that racism and classism, as well as other oppressions that often negatively affect men to a degree not always less than women, are rooted in patriarchy (Johnson 2014, 71).

4. It seems that Marilyn Frye's words written in 1983 are also relevant today. Frye wrote about women's oppression, which is universal. It involves oppression in terms of function, which is to serve men. The nature, course, and specifics of this serving may vary locally and look different depending on the class and race of the woman, as well as other parameters. Nevertheless, it is women who serve men in many areas ("service sector") rather than men who serve women (Frye 1983, 9–10; see also Brittan and Maynard 1984). It is also worth noting the connection between misogyny and sexism. The servility of women toward men has led to the fact that women have been, and often are to this day, treated as persons only relationally, that is, in relation to some man for whom a woman is a mother, sister, grandmother, wife, etc., but not a person as a person. Her personality is owned and appropriated by a man (Manne 2018; Kukla 2020).

5. Anti-abortion legislation is primarily a type of control of women and an expression of the power of patriarchy, not a concern for life. Interestingly, the term "fetus protection" has been imposed by abortion opponents as the dominant form of abortion discourse (Kolbert and Kay 2021, 7). Like many other actions and norms imposed by men on women, abortion control is men's control over women's bodies in order to strip them of their autonomy. Carole J. Sheffield refers to this patriarchal policy of constant intimidation of women by men as "sexual terrorism" (Sheffield 1995, 1). Anti-abortion legislation in the United States is not as strict as in Poland, as the *New York Times* warns, pointing to the negative example of Poland (Bennhold and Pronczuk 2022).

6. Maguire states that one should not extrapolate the point of view of a particular group of women and assume that a particular empowering situation functions as empowering for all women—it may be exactly the opposite, as in the case of reproductive technologies. Therefore, the essence of feminism, despite differences in perspectives, remains the struggle against sexism and injustice (Maguire 2019, 226–227).

7. For noteworthy attempts to combine liberalism with the ethics of care and the category of dependency, see Bhandary and Baehr 2021.

8. The strand of feminism I present can be defined in the footsteps of Cinzia Arruzza, Tithi Bhattacharya, and Nancy Fraser as "feminism for the 99%," which is definitely not an elitist liberal feminism that privileges only the 1 percent. It is a feminism that is concerned with the welfare of the poorest and most excluded, negatively experienced by the capitalist appropriation of land, raw materials, and labor, additionally today suffering in particular from climate change. The anti-capitalist and anti-racist profile of this feminism is significant (Arruzza et al. 2019). On the feminist critique of liberalism, see also Schwartzman 2006.

9. Etymologically and genealogically, feminism is focused on the struggle against the subordination of women. However, because women are also subjugated on the basis of their race and/or class, feminism also extends to racism and classism. Feminism often goes beyond fighting sexism because of its commitment to egalitarianism (Stone 2007, 204). I share the view of those feminists who believe that the concept of oppression is applicable only to women, who are structurally oppressed solely because they are women. Men are not oppressed, while the fact that they may often be in a bad situation needs to be described using other, more appropriate terms, but not the term "oppression." See, among others, Frye 1983, 1–16.

10. I use the term "race" in the sense of a social category, which, as Joseph L. Graves Jr. and Alan H. Goodman point out, is the erroneous effect of ideas about ancestry and appearance (Graves and Goodman 2022, 3). Nevertheless, as Charles W. Mills points out, even as a sociopolitical category, the concept of race in practice is no less real than if it were a biological category (1997, 126). Racism, on the other hand, in addition to prejudice, requires first and foremost an ideology that makes racism a compact, integrated system, as opposed to individual prejudices, which can be random and lack structural translation (Bonilla-Silva 2022, 25; see also Zack 2017).

11. The category of gender is debated; among other things, the paradigm of universality (realism) is accused of understanding womanness as if it were a Platonic idea (see Mikkola 2016, 28–37). No uniform definition of womanhood can be established, but this does not change the fact that female human beings have been, and continue to be, dominated and marginalized by male human beings. The understanding of gender in this book is performative (Mikkola 2016, 36).

12. This is also evident in the masculinist bias exposed by feminist criticism of the history of philosophy (see Alanen and Witt 2004).

13. The idea of a mother and motherhood is still presented in an idealized way, assuming a certain degree of self-sacrifice on the part of the mother who devotes herself to the happiness of her child, who has the highest value. This essentialist understanding of motherhood may be reinforced by the instability and stressful nature of capitalism (Teodorescu 2018, 78–82, 91). Another consequence of the motherhood category is the expectation of sacrifice on the part of female mothers, as well as blaming mainly women for any parenting failures (Eyer 1996). Another of the consequences of relegating everything related to women to the private sphere has become the lack of presence in the culture of positive, common categories expressing the needs and interests of women. This is not about the image of a mother or parent, but a positive category. Also, their corporeality was not affirmed in the culture, which could have consequences for their health (Rawlinson 2001, 411–412). In the long run, this could also have had negative consequences for their status in bioethics.

14. This section is a slightly modified and shortened version of Szocik 2022.

15. For a detailed overview of the issues and theories analyzed within the feminist approach to science, as well as the feminist philosophy of science, see Bleier 1986; Harding and Hintikka 2003; Crasnow 2020; and Anderson 2020.

16. A particular challenge to feminism is sexual selection theory, where the data confirm invariant sexual preferences consistent with biological adaptation theory, i.e., women's preference for older men with resources, as well as men's preference for younger, sexually attractive women (Gray and Garcia 2013, 45).

17. It is worth mentioning here one of the precursor feminist science fiction novels, *The Left Hand of Darkness*, published by Ursula K. Le Guin in 1969, in which the author deals with androgyny to dissect power structures and show a vision of the world without rigid gender roles thanks to hermaphrodites (Le Guin 1969).

18. Susan Gelfand Malka shows that nurses in the United States have often tried to combine the ideas of second-wave feminism with a sense of a kind of calling to care and responsibility for patients, remaining the only party in the American business model of healthcare interested in patient welfare from the perspective of an ethic of care (Malka 2007, 167–169).

19. Socialist feminism shows that capitalism goes hand in hand with patriarchy in oppressing and dominating women. The problem is not the unequal competition for access to free market resources between men and women; the problem is negative attitudes toward women and sexual classism.

20. An interesting defense of feminist political liberalism by Lori Watson and Christie Hartley is worth reading. The authors equate feminist liberalism with political liberalism, which in turn is understood as the defense of the interests of the person understood as a free and equal citizen, and which also opposes gender elements that limit individual freedom and interests (Watson and Hartley 2018).

21. The category of relationality is central to feminist bioethics, as it allows analysis of as many factors as possible that influence an individual's decision (Sherwin and Stockdale 2017).

22. It was not until the nineteenth century, especially in the United States, that men replaced women midwives and general healers with male physicians, pushing women out of the medical profession. This resulted in a very strong stereotyping of the medical and nursing professions until today. This sexism in medicine has a definite economic motive, although it is not necessarily motivated solely by the pursuit of a financial monopoly (see Ehrenreich and English 1973).

23. Feminist principles applied to decision-making also include, but are not limited to, consideration of women's specific experience, pluralism, egalitarianism, the importance of relationships, awareness of harmful environments, the impact of external factors on the individual, and skepticism about traditional health policies (Ballou 1995, 44).

24. An example of a decidedly nonfeminist approach to bioethics is the Kantian and subsequently Kantian-modeled understanding of moral subjects as equal despite obvious differences between them, such as in the situation of organ selling (MacDougall 2022, 167).

25. Beauchamp and Childress list some of the feminist theories as examples of the use of bottom-up methodology in bioethics. This is certainly true for feminism per se, but it is difficult to agree with their further statement that bottom-up approaches do not offer permanent normative conclusions, only provisional ones, because they are contextual and evolve over time (Beauchamp and Childress 2013, 398). Feminist bioethics is guided by the principle of respect and equality for the excluded, and it is difficult to imagine a context in which this principle would be abolished.

26. Diana Tietjens Meyers discusses the problematic nature of female autonomy complicated by cultural and social discourse on female reproductive biology (Meyers 2002).

27. The limitation of autonomy and lack of equality with men are especially true for pregnant women, due in part to the lack of a pregnant male figure or equivalent to be protected by law (Kendal 2018, 63), and then this legal status was extended to pregnant women, as has happened to women in other areas as a result of feminist pressure.

28. The concept of relational autonomy is further reinforced by the processual philosophy of biology, which draws attention to the ecological interdependence between organisms. Living organisms are relational entities, not fixed Aristotelian substances, and constantly depend on their environment (Dupré and Nicholson 2018, 20–21).

29. "Masculine" normative social contract theories did not take women into account, as Carole Pateman argues in her concept of the sexual contract (Pateman 1988). Charles W. Mills, on the other hand, makes an even more profound unmasking of the social contract, which was primarily a racial contract, because it proposed different rules, values, and norms for whites and nonwhites (Mills 1997). Therefore, as Mills argues, for nonwhites, as opposed to whites, the philosophical differences between normative systems played little role, as they were abstract, ideal theories. What mattered to nonwhites was the involvement of the racial contract, which set the framework for normative theories (Mills 1997, 110). Similarly, Mills criticized Western liberalism, which he interpreted in practice as racial liberalism, in turn denounced by nonliberal feminists as patriarchal liberalism (Mills 2017).

30. See also Tännsjö 1998 on the similarities between utilitarianism and feminism.

2

An Introduction to Feminist Bioethics in Space

Methodology

Introduction

In the understanding of space bioethics that we have proposed (Szocik 2023), the distinctiveness of the space environment from that of Earth plays an important role. This difference justifies, at least to some extent, a different treatment of principles and rules than that to which we have been accustomed by nonfeminist bioethics on Earth. Thus, as we have suggested, the context and circumstances justify a different application of bioethical norms and allow for a different evaluation of the same biomedical procedures in space and on Earth. The mere fact of exposing the role of context and circumstance does not yet establish the feminist character of a given approach. Nevertheless, the methodology of feminist bioethics exposes the importance of the details and context of a given situation (Sherwin 1996, 51).

In order for a reference to context and detail to receive the label "feminist," it cannot be disposable and only apply to different backgrounds. It is therefore not enough to say that a certain set of principles and rules apply according to certain patterns on Earth and according to others in space, but that within these environments regularities and sequences of moral decisions are maintained. In order to make the approach that exposes the importance of context and detail feminist it is important to realize the uniqueness and singularity of each situation in a given environment. Thus, we can say that for feminist space bioethics, no situation concerning the application of, for example, human enhancement for space missions will be the same. The moral judgment of applying human enhancement to each astronaut may be assessed differently even if there is seemingly nothing different between them and even if all astronauts in a given rated group are to receive the same biomodification and participate in the same mission. Thus, for feminist bioethics of space exploration, contextualization means not so much the new environment—such

Feminist Bioethics in Space. Konrad Szocik, Oxford University Press. © Oxford University Press 2024.
DOI: 10.1093/9780197691076.003.0003

is the importance of contextualization for nonfeminist bioethics of space missions—but the conditions of the individual.

What distinguishes feminist from nonfeminist space bioethics is the analysis of particular biomedical issues in terms of oppression and exclusion. Feminist space bioethics will therefore, like nonfeminist space mission bioethics, analyze the legitimacy of, and justification for, among other things, human enhancement, gene editing, moral bioenhancement, and other procedures in space. Unlike nonfeminist bioethics, a feminist perspective asks who gains and who loses from applying or not applying a given procedure, the position of the various parties involved in a given procedure, and the degree of their freedom, autonomy, power, independence, and interdependence. An example of a feminist bioethics of space exploration would be the question whether genetically modified individuals sent into space would be exploited for the sake of the wealth of states and corporations. However, the feminist bioethics of space exploration does not focus solely on the analysis of biomedical procedures, but on the entire social and political context and the many levels of sexism, racism, and other forms of oppression in which the individual operates.

In a sense, feminist bioethics of space missions is doubly different from nonfeminist bioethics. The first difference concerns the aforementioned feminist perspective, which introduces categories and basic concepts specific to feminist philosophy. The second difference is in looking at the intrinsically oppressive potential of the space environment that bioethics focuses on. This is the specific environment of space missions, which has some similarities to specific environments on Earth, extreme environments such as the battlefield or Arctic expeditions. However, it is only the feminist perspective that draws attention to the potential oppressiveness of this environment in terms of gender, race, or disability. Even when other work indicates that the space environment will be extreme, analyses of the problem are usually devoid of reference to oppression against gender or race (Cockell 2016, 2022). This is one of the questions posed in this book: whether oppression in space is different from oppression on Earth, and whether exclusion and power structures will function differently than on Earth.

The adjective "feminist" added to the term "bioethics" is problematic in a specific way, perhaps more so than other terms for bioethics such as "deontic" or "utilitarian," for example. The term "feminist" does not mean just another normative theory. The core of the term suggests biological sex. Is it correct, then, to assume that nonfeminist bioethical concepts can be equated with what is masculine? This is the dominant feminist view of the history of philosophy, which is masculinist and usually anti-woman. But as with feminist philosophy, feminist bioethics is more justified in emphasizing new insights into old problems, as well as in exposing the significance of problems previously overlooked by nonfeminist bioethics and not recognized as a problem (see Jean Grimshaw on the novelty introduced by feminist philosophy, which should not be read in terms of a search for a specifically feminine or feminist element, but rather in a new look at philosophical problems and answers) (Grimshaw 1986, 259–261).

The following understanding of feminist bioethics seems particularly pertinent:

> A feminist bioethics, then, paying attention to sexual difference, the body, and material forms of life, advances a political and ethical program that focuses, not on the equality of rights, but on *women's* rights. Rather than assuming that the rights articulated in the name of "man" are adequate to describe human rights, a feminist bioethics analyzes women's experience and women's bodies to articulate those civil rights discovered through this other access to universality. It seeks to produce new figures of subjectivity and intersubjectivity and new ways of thinking universality based on analogies and metaphors drawn from women's experience and bodies. It requires all those cultural practices in which we represent and realize ourselves to begin again from this other body, thinking it not as an aberration or exception, but as another way to approach *human* rights. Feminist bioethics seeks to address the attenuation of the human in "man," as well as the very real illness and suffering that attends it. (Rawlinson 2001, 414)

Feminist bioethics in space applies categories of feminist bioethics such as analysis of power dynamics, and issues of exclusion and marginalization, as well as injustice (Rogers et al. 2022, 2), that may arise in bioethical situations during space missions.[1] And in keeping with Rawlinson's quoted understanding, feminist bioethics in space is looking for what is specific to the female perspective, expanded to include the perspectives of other traditionally marginalized groups.

Finally, it is worth noting that just as mainstream bioethics is devoid of feminist categories,[2] analogously the discourse on space is, with few exceptions, devoid of a feminist perspective. Despite the belief of some theorists about the end of feminism and the postfeminist era, outer space has not yet experienced the cultural and ideological period of feminism. It is important for humanity, as Francesca Ferrando contends, to get rid of dualistic thinking patterns in classifying people and the environment before embarking on intensive space exploration (Ferrando 2016, 138). Feminism is a tool for getting rid of these patterns, as well as drawing our attention to their existence.

Autonomy in Space

The space mission environment poses a particular challenge to the principle of respect for autonomy and the rule of informed consent, although there are some parallels to military ethics. Nonfeminist ethics, especially that based on the four principles of biomedical ethics, finds the situation of the astronaut under pressure quite conceptually challenging and difficult to characterize unambiguously (Szocik 2023). It seems that from the point of view of nonfeminist bioethics, for an astronaut giving informed consent to participate in research on the International Space Station, in which she does not want to participate but fears the consequences of losing her contracts in the future, as long as she has given her consent, she is autonomous. Feminist bioethics proposes here the notion of relational autonomy discussed in chapter 1. According to this perspective, this astronaut, like any other human being, has never been, is not, and never will be autonomous in the way that being autonomous is understood by nonfeminist bioethics. The high level of competence required by the astronaut profession, the enormous costs of both preparing for and executing a mission, and the social, political, but also scientific expectations of the astronaut mean that any decision will be immersed in these multiple relationships and dependencies. The concept of autonomy in a nonfeminist sense lacks utility in this context.

This feminist conception of relational autonomy is an offshoot of relational theory, which underpins all types of relationships and values that characterize bioethical situations. Feminism points out that our autonomy is often constrained, meaning that even despite the theoretical possibility of choice, our choices in practice are either impossible or limited. One could wonder if the limitation of autonomy in space is greater than it is on Earth or simply different. The specifics of the mission and the infrastructure, as well as the dependence on the life support system, will make it impossible to change the environment or the community. For example, a woman interested in getting an abortion where it is forbidden will not be able to go to another country that allows abortions or performs them illegally. In this specific infrastructural and environmental sense, the space mission environment further limits autonomy, which is already restricted in analogous situations on Earth. In space, however, there is no alternative, which at least theoretically exists on Earth, however difficult it may be to access.[3] A woman interested in having an abortion whether on Earth or in space may have limited autonomy due to a number of factors, but it seems that the space environment may compound these multiple constraints and make a woman interested in having an abortion even more vulnerable and even less autonomous than on Earth.

Thus, it can be said that in a sense, the environment of space missions does not limit autonomy any more than various situations on Earth as long as our autonomy is always limited, as indicated by the feminist concept of relational autonomy. Adopting a feminist conception of autonomy results in those situations that nonfeminist bioethics interprets as autonomous not being so. On the other hand, this does not change the fact that the environment of space missions may foster additional limitations on autonomy with regard to certain biomedical situations. Such situations could include reproduction issues in space, under the hypothetical assumption that this would not only be possible, but also desirable. As will be discussed in chapters 5 and 6, reproduction in space may be subject to specific regulations and restrictions that will primarily affect women and may specifically limit their autonomy.

Justice in Space

Another key concept in bioethics is justice. Feminist bioethics points out that the dominant nonfeminist understanding of the principle of justice is abstract and ahistorical. This is because this nonfeminist understanding of justice focuses on the individual cases rather than structural, historically grounded injustice (Fourie 2022, 30–31).

The principle of justice in relation to the space mission environment will apply differently to individual astronauts depending on their gender and race, as well as other categories. There is no doubt that the basic dividing line with regard to biomedical situations in the context of the principle of justice will run between men and women. There is a considerable risk that just because a participant in future space missions is a woman, she will be subject to different rules and criteria, offered different—probably narrowed—opportunities, and burdened with additional responsibilities compared to a man. This may be particularly true for reproductive rights in the hypothetical scenario of human reproduction in space, where there may not be equity between women and men in this regard. Injustice can in practice affect everyone who is not a gender binary man. It is worth remembering, however, that injustice can manifest itself in the form of a specific combination of different types of oppression and exclusion caused by overlapping categories such as gender, age, class, or race (Collins 2015; Crenshaw 1991).

The perspective taken by postcolonial and multiracial feminism is important for capturing the feminist bioethics of space exploration presented in this book. It relies on not adopting the Western perspective inherent to the white, historically privileged as the default perspective (Friedan 1963), which led to the fact that for nonwhite feminists, white leaders of the feminist movement began to be seen as a privileged class, imposing their approach to seeing the world in a way analogous to the dominant male point of view taken as default and universal (Newman 1999; Lugones 2003; Whipps 2017, 24–25). Space exploration, which requires advanced and expensive technology and access to resources, has the potential to focus on ethical and bioethical issues inherent to the privileged inhabitants of Earth. Therefore, a postcolonial and multiracial perspective is particularly relevant for considering bioethical problems in such a specific environment as space. Expanding on postcolonial feminism, Namita Goswami shows how the Eurocentric paradigm of homogeneity essentially threatens the entire planet by negating the idea of heterogeneity (Goswami 2019). Multiracial feminism, on the other hand, shows that race is a very powerful factor of oppression in all dimensions and, according to the concept of intersectionality, specifically causes the oppression of women of color (Wing 2003). The problem of racism can be a particularly significant barrier to just selection of astronauts in those countries where racism operates analogously to casteism (Wilkerson 2020).[4]

Is Feminist Bioethics of Space Exploration Different from Feminist Bioethics on Earth?

Even after outlining the potential for limiting autonomy and inequitable structures in space, it is still difficult to say whether the space environment is different enough from environments on Earth, including extreme environments, to warrant a different approach to bioethics than on Earth. If we are talking about nonfeminist bioethics, it is certainly possible to point out some differences that make space mission bioethics somewhat similar to military bioethics and bioethics of extreme environments (Szocik 2023). On the grounds of nonfeminist bioethics, we are usually inclined to treat biomedical situations under normal, standard conditions differently from those under abnormal, extraordinary conditions. Usually the latter means minimizing the rights of the individual and being more inclined to focus on the success of either the group or the task. To the same or a similar degree, an approach that minimizes individual rights can be present in military ethics and bioethics, as well as in the bioethics of space missions.

Such a differential approach in bioethics depending on the situation is possible on the grounds of nonfeminist bioethics, which is based on abstract principles. Feminist bioethics, on the other hand, takes into account the complexity of an individual's situation. This situation is always complex, depending on the context, the categories assigned by society, and the networks of dependency. Among the basic categories discussed by feminist bioethics are sex and gender, as well as exclusion and submission. Regardless of the environment, earthly or cosmic, an individual belongs to a specific category such as sex and gender, class, race, and age group, each of which has a certain degree of power and number of opportunities, as well as being discriminated against or itself participating in the discrimination and exclusion of others. Therefore, from this point of view, it can be said that differences in the evaluation of biomedical situations in different environments, earthly and cosmic, in the light of feminist bioethics will either not occur at all or will be relatively small, in contrast to the perspective of nonfeminist bioethics, which tends to differentiate the ethical evaluation of a given situation based on its context, task, and purpose.

Thus, this is a methodological peculiarity of feminist bioethics, which always analyzes a specific biomedical situation, taking into account the complexity of the individual's situation and her belonging to different categories, as well as her possibilities in terms of power and being excluded, or power to exclude others.

On the other hand, there may be certain scenarios in future long-term space exploration that could be specifically discriminatory and oppressive, in ways perhaps unparalleled in any Earth environment. However, this does not mean that feminist bioethics itself should be modified to remain capable of adequately and effectively analyzing and resolving moral dilemmas in such challenging situations. The differences, therefore, relate not so much to the bioethical principles and norms themselves, as to the degree and nature of the oppressiveness of different environments. It is not entirely certain that even the most oppressive scenario imaginable in future space exploration would be qualitatively different from situations familiar from Earth in which women, racial and ethnic minorities, sexual minorities, or other marginalized groups have been subjected to oppression. However, this does not change the fact that the space environment can be particularly oppressive for certain groups such as women or people with disabilities.

The value of feminist ethics and bioethics applied to the moral analysis of space missions is in defending certain values, such as women's autonomy, as well as in criticizing certain practices, such as the exploitation and subjugation of women, regardless of place and time. If some feminists criticize sex work as well as pornography, the same feminists should criticize sex work and pornography in space if they were to occur there. This is a kind of inflexibility of feminist bioethics that reveals its commitment to certain principles, perhaps to a greater degree than, for example, to caring. If a practice is morally wrong on Earth, such as the enslavement of racialized minorities or the oppression of women, feminism will not justify it as morally good or neutral in space just because of harsh environmental conditions. This specifically functioning inflexibility is different from the inflexibility that characterizes nonfeminist bioethics, criticized by feminism precisely for its attachment to abstract and general principles. Feminism's inflexibility refers to disagreement with the violation and minimization of the validity of human rights and norms protecting in particular those vulnerable to exploitation and abuse. In this respect, perhaps somewhat paradoxically, feminist bioethics of space exploration proves to be inflexible compared to flexible nonfeminist bioethics.

From the point of view of feminist bioethics, the space mission environment does not justify exceptionalism in the application of bioethical principles and rules. Gender, power structures, and mechanisms of exclusion must always and everywhere be kept in mind. Mainstream bioethics, on the other hand, is open to the rule of moral exceptionalism. Therefore, nonfeminist bioethics may be inclined to highlight parallels between the military/battlefield environment and military bioethics and the space environment and space mission bioethics. Pointing out such parallels leads to an acceptance of the suspension of certain norms, an acceptance of paternalism. An example of such a critique of the idea of the moral uniqueness of the space environment, albeit not undertaken on feminist grounds, is the approach of Maurizio Balistreri and Steven Umbrello (2022). Various biomedical procedures that are controversial in nature, such as human enhancement, should always be evaluated in the same way, regardless of the environment, context, and purpose. The bioethics of space missions combines the idea of inflexibility with particularity and contextuality. Feminist concern for the weakest and excluded provides a special justification in favor of protecting the rights of the weakest in such a particularly challenging environment, where the risk of violating norms may be greater and acquiescence to such violations may gain approval and understanding. In this specific sense, feminist bioethics of space exploration is context-sensitive, which, however, unlike nonfeminist bioethics, serves to oppose the minimization or suspension of certain norms, rather than to justify the modification of bioethical norms. The idea of moral exceptionalism has historically been harmful and used to justify antinatalism (in its negative political sense, known from history, and not the philosophical sense we discuss in Chapter 6), eugenics, and sterilization. In contrast, feminism's proper reference to particularity and context is preserved in feminist bioethics of space missions. This is because feminism turns its attention to justifying human exploration and exploitation of space. As space exploration (understood as a human or uncrewed mission aimed at scientific exploration and eventual human settlement in space) and exploitation (denoting profit-oriented missions, mainly the extraction of raw materials and land) are largely dominated and motivated by nationalism, capitalism, and militarism, the justification for controversial biomedical procedures decreases rather than increases. For feminism, therefore, the full context is not just harmful conditions in space, but precisely these motivations for space missions.[5]

Long-Term Space Missions

Feminist bioethics of space exploration can be applied to all biomedical situations that take place during space exploitation and exploration involving humans. For the purposes of this book, the most useful and also the most basic division of space missions is between short-term and long-term missions. Short-term missions are those currently on the International Space Station, lasting several months. Location is also an important criterion. Short-term missions do not go beyond Earth orbit. In contrast, long-term missions, as defined in this book, are those lasting at least several dozen months and carried out beyond Earth orbit. The main example of such a mission would be a mission to Mars. These are missions that humanity has not yet carried out. In contrast, bioethical challenges in short-term missions, so far carried out in orbit, are also important, while resembling considerations of clinical problems discussed on Earth (Sawin 2021).

The subject of interest in this book is the long-term missions that humanity is yet to carry out in the future. Such missions can be carried out for scientific or commercial (nonscientific) purposes, as well as with the idea of establishing a permanent habitat for humanity in space (the space refuge concept) (Billings 2021; Changela et al. 2021). To some, such a classification may seem a bit too idealistic, as political pragmatism may suggest that a presence in space requires a strong political motivation, and such usually goes hand in hand with nationalism, militarism, and capitalism.

Regardless of the actual motivation, however, the book assumes that humanity will carry out long-term space missions beyond Earth orbit in the future. The bioethical challenges that characterize such missions will differ from those inherent in short-term missions in Earth orbit. The two main challenges are human enhancement and human reproduction in space. It is around these issues that the feminist bioethics of space presented in this book is developed.

Feminist Bioethics of Space Exploration and the Philosophy of Space Missions

It is worth keeping in mind what space exploration bioethics is, and remembering its distinctiveness from the philosophy and ethics of space exploration, as well as space policy. One definition of bioethics is as follows:

> Bioethics is applied ethics focused on health care, medical science, and medical technology. . . . Ranging far and wide, bioethics seeks answers to a vast array of tough ethical questions: Is abortion ever morally permissible? Is a woman justified in having an abortion if prenatal genetic testing reveals that her fetus has a developmental defect? (Vaughn 2023, 4–5)

Feminist bioethics, in turn, offers its own specific methodology in response to this biomedical issue:

> The value of feminist bioethics lies in more than just its ability to focus on healthcare topics and practices that are of particular relevance to women. By paying attention to power dynamics, epistemic biases and the particularities that cause certain groups to be excluded and marginalized, feminist perspectives offer distinctive tools for analyzing all situations of social injustice. As a result, feminist bioethics provides insight into forms of harm and injustice that mainstream bioethics routinely misses. It is this capacity that has helped build feminist bioethics' strong intersectional connections with disciplines such as disability bioethics, Black bioethics and postcolonial bioethics. (Rogers et al. 2022, 2)

Bioethics is thus distinguished from philosophy and ethics, as well as space policy, which are interested in analyzing nonbiomedical issues. A feminist perspective, however, not only offers new explanatory categories but rejects previous nonfeminist perceptions and explanations of reality. For this reason, however feminist bioethics distinguishes itself from other disciplines, references to nonbioethical contexts will appear in this book. As long as feminism pays attention to exclusion, oppression, and unjust power structures, feminist bioethics will require reference to a broader, nonbioethical context. If feminist bioethics draws attention to the unequal situation of black people in healthcare in the United States, it is difficult not to relate their inferiority in healthcare to extra-health structures, beginning with their exploitation as slaves, which underpinned US statehood. An analogy should be made to structures outside the healthcare system if one analyzes the inferior position of women within feminist bioethics, even if only their exclusion from clinical trials. It is impossible not to refer here to the stereotype of women as mothers, which affects women's opportunities and possibilities in many areas of life, not only healthcare, but also education and work.

Similarly, in the case of feminist bioethics of space missions, it is difficult not to refer to the broader, nonbioethical context. When analyzing the ethical status of the concept of radical human enhancement for space missions, it is important to pay attention to the dynamics of power structures, who benefits and who suffers losses and is exploited in these structures. Thus, this applies not only to the human enhancement procedure itself, but to the justification for space missions. This is a peculiar feature of feminist bioethics, which, even if conceptually narrowed to biomedical issues in accordance with the meaning of the term "bioethics," pays attention to the status and condition of people in a given biomedical situation in a broader, nonbiomedical context before analyzing it. For this reason, this book also discusses issues and contexts that are not directly related to bioethics but show the context in which the bioethics of space exploration will operate.

Feminist bioethics of space missions, as well as feminist philosophy and ethics of space exploration, despite their critical and skeptical approach to space exploration, do not oppose the very idea of space exploration, understood as the search for the next new place for humanity beyond Earth's borders. In this respect, they are similar to the ethics of space exploration in the way of understanding it proposed by Tony Milligan and J. S. Johnson-Schwartz. The authors emphasize that the task of space ethics is not to stop or control space exploration, but to morally evaluate this important type of human activity, which contains risks and dangers in addition to positive elements. The following quotation illustrates perfectly the task and nature of space ethics:

Space ethics poses various challenges and dilemmas, as well as opportunities. . . . To ignore the challenges and dilemmas is to be poorly prepared. It is a little like trying to go to Mars before the proper technology is in place, and without any means of return—a proposal that has been made more than once. Such an approach would itself pose numerous ethical problems. Preparation for future human activities in space, as on Earth, is a many-sided endeavour, and ethics is part of the preparatory process. (Milligan and Johnson-Schwartz 2023, 112)

The understanding of feminist bioethics proposed in this book, as well as the feminist approach to space exploration in general, presupposes such an understanding of the bioethical and ethical assessment of space exploration, in which human presence in space is treated as an opportunity and perhaps even a necessity, but with full awareness of various risks.

Finally, it is also worth recalling the role played by futures studies, of which this book is a part. Much attention has been devoted to the connection between space bioethics and futures studies (Szocik 2023). Bioethicists consider the role played by future technologies even if they are not yet available today. We sympathize with the approach presented by John K. Davis analyzing the ethics of life extension. Although these technologies in the sense described by Davis are not available today, the author analyzes the potential consequences of their application in the future such as the risk of overpopulation and the issue of equal access to these biomedical technologies (Davis 2018). Similarly, some of the issues considered in this book require advanced space technology and medicine, such as the possibility of reproduction in space. However, it is worthwhile analyzing such issues in order to know in advance whether this is the type of future for humanity that we might consider preferable.

Methodology

Introduction

From what has been said so far about feminism, it is clear that feminism is primarily a new methodological approach. Feminist bioethics should not only analyze the classical problems of nonfeminist bioethics using feminist categories but also deal with new topics that are overlooked by canonical bioethics. A feature of nonfeminist bioethics is its focus on bioethical problems specific to privileged people who have access to advanced medical care—hence the domination of bioethics by issues that are typical of medical ethics, which are dominated by advanced medical technology, with human enhancement and reproductive technologies at the forefront as popular topics. If feminist bioethics is treated, as Arianne Shahvisi suggests, as part of anticolonial bioethics, then not only the methodology but also the problems analyzed should reflect the bioethical problems of the excluded and marginalized (Shahvisi 2022). Western-centered bioethics needs to be decolonized, which means adopting the perspective of representatives of non-Western—or more broadly, non–Global North—viewpoints, and being aware of "the racialized, gendered and geographical patterns of suffering" (Shahvisi 2022, 210).

The feminist bioethics of space exploration is somewhat of an exception. Due to the nature of the environment of space missions, which must be highly technologically advanced, the dominant themes in this book concern feminist approaches to human enhancement and reproductive technologies. Such is the nature of long-term and deep-space missions, which would not be possible without constant monitoring of the human body, digital health, possibly radical human enhancement, and assisted reproductive technologies (ARTs)—if ever human reproduction in space were to be possible. In this context of a high-tech space exploration environment, it is worth bearing in mind cyberfeminism, which points to the technological exclusion of women. Because of patriarchy and sexism's inherent association of women with the private sphere, women were excluded from participation in public life, including in science and technology. The understanding of technology is also masculinized (Genz and Brabon 2009, 147).[6] In the arguments of US opponents of including women in space missions, one of the arguments was the alleged ignorance of women regarding the use of technology. On the other hand, the inclusion of women in the world of technology brings with it as one of the negative byproducts increased monitoring of the female body through new technologies (Turkmendag 2022).

The feminist bioethics of space exploration as presented in this book deals with topics that feminist bioethics often criticizes as topics and problems for the privileged. However, we assume that human enhancement and reproduction will be the main bioethical challenges during future long-term space missions. Therefore, the specific environment to which feminist bioethics is applied affects its thematic scope.

The Problem of Principles in Feminist Ethics and Bioethics

As we presented in the previous chapter, principles and rules are the basis of nonfeminist bioethics. Do feminist ethics and bioethics recognize the existence of any principles to which they attribute the character of, if not absolute, then at least universally accepted starting premises? This is a challenge within feminist methodology insofar as feminism emphasizes the dominant role of standpoints (Hutchison 2022, 45), stresses the relativity of location, and above all criticizes nonfeminist ethics and bioethics for the cult of abstract, general principles and rules (Code et al. 1988).

There is no doubt, however, that feminism recognizes the existence of at least a few principles that could perhaps even be considered in terms of absolute principles. By pointing out that women are worse off than men, feminism endorses the principle of beneficence of women and all oppressed people. By emphasizing that exclusion and discrimination of women and other groups is unjust, it exposes the role played by justice and fairness. Moreover, at least liberal and libertarian feminism stresses the necessity of establishing equal rights and opportunities, and the postulate of the principle of equal consideration of interests of men and women is based on the principle of justice. Feminism based on the ethics of care particularly emphasizes the role of the principle of beneficence. It is therefore legitimate to translate the main ideas of feminism into the language of classical bioethical principles. However, it is worth remembering that the central element of feminist philosophy and bioethics is the achievement of gender equality, as well as the fight against all types of social inequality, so the principle of social justice can be considered a universal principle of feminist bioethics.

The methodology of feminist ethics and bioethics is no less principle based than classical principlism and normative ethics such as utilitarianism and deontology. Rather, the difference is in how we think about principles and the criteria for applying them. Feminism offers its own version of the reflective equilibrium method, in which the specification and weighing of methods and the balancing of principles and rules take place in terms of relationality, personal involvement, sympathy, and empathy, and undoubtedly with a greater involvement of a wide range of different emotions. This is not, however, a classical or even modified understanding of the reflective equilibrium method, in which the starting point is to address our moral intuitions and moral judgments, and then dynamically use mid-level principles (Arras 2009; DeGrazia and Millum 2021).

If we take power as the key category, what is important for feminism is granting women, and women's vision of morality and their moral reasoning, power. That is, when analyzing the same biomedical situation, both a woman and a man can come to the same conclusions, but what is essential is that the woman's vision of the biomedical situation in question be given voice. There is no rule that, when making the decision to stop a therapy will inevitably end in the patient's death, only a woman will defend the patient's right to life and stand in solidarity with the family, while a man will always abstractly compare principles and count costs. The opposite situation may occur. What is important is that the woman's point of view, whatever conclusions it may lead to, should not be marginalized or treated worse or with distaste or pity through the prism of gender stereotypes. Giving priority to the female perspective, however, is not so obvious and clear-cut. Concepts such as false consciousness and adaptive preferences illustrate that women can be part of a system of oppression and discrimination no less than men. This is particularly evident in a system of liberal individualism, where women, socially and professionally equal to men, can contribute equally to the maintenance of a patriarchal system of oppression.

The difference in methodology between feminist and nonfeminist ethics does not, therefore, lie in different sets of principles but in different criteria for their application. The different criteria, in turn, stem from feminism's emphasis on the contextuality of the individual. However, one cannot say that nonfeminist bioethics does not take contextuality into account. This is not the kind of contextuality that characterizes feminism—let us call it the rule of absolute contextuality. The contextuality that characterizes nonfeminist bioethics may be called "relative" or "graded" contextuality. While Tong's charge against traditional ethics is that it assumes the absolute autonomy of the moral agent and the universality and impartiality of principles (Tong 1993, 11), this charge cannot be applied to contemporary bioethics. In bioethics, the gradation of autonomy is acknowledged, which is expressed by much attention being paid to the rule of informed consent, which has its roots, among other things, in the principle of autonomy. Neither can contemporary bioethics be accused of recognizing principles as absolute and impartial. This contradicts the notion of reflective equilibrium, the main idea of which is the conviction that just at the starting point we never know which principle we will reach in a given situation. What we can accuse nonfeminist contemporary bioethics of is not paying enough attention to the specific situation of excluded and marginalized groups. Feminists in both ethics and bioethics point to the objective basis of their criticism of the male point of view in moral justification (Jaggar 2000).

The relationship of feminist bioethics to principles can also be understood differently. Lewis Vaughn (2023, 52) believes that feminists are not so much interested in principles and bioethical theories as in the social realities in which decision-making individuals—mainly women—function.

In conclusion, under certain conditions, it can be said that feminist bioethics respects the same principles that nonfeminist bioethics does. However, feminist bioethics introduces, at the starting point, categories essentially unseen by nonfeminist bioethics, namely sex and gender on the one hand, and the categories of exclusion, oppression, and power structures on the other (Table 2.1).

Feminist Critique of Nonfeminist Bioethics Methodology

Despite outlining a kind of methodological affinity between feminist and nonfeminist bioethics with regard to thinking about principles, feminism subjects nonfeminist bioethics to radical criticism. One of the pillars of nonfeminist

Table 2.1 A comparison of nonfeminist and feminist bioethics of space exploration

Nonfeminist bioethics of space exploration	Feminist bioethics of space exploration
Different methodology due to environmental differences	Same methodology regardless of environmental differences
Explaining biomedical situations in terms of principles and rules	Explaining biomedical situations in terms of sex, gender, oppression, exploitation, and others appropriate to feminism
The tendency to compare the moral ecology of space to the moral ecology of other extreme environments, especially military ethics and the battlefield	Reluctance to compare the moral ecology of the cosmos to the moral ecology of other extreme environments
Willingness to morally justify biomedical procedures controversial on Earth	Reluctance to morally justify biomedical procedures controversial on Earth
Nearly unanimous justifications for applying radical bioenhancement	Skeptical and cautious approach to the idea of radical bioenhancement
Uncritical look at space policy and nonbiomedical context	A highly critical look at space policy and the nonbiomedical context, which may even call into question the viability of missions as well as biomedical procedures in space
Easy justification for excluding people with disabilities from space missions or possibly subjecting them to mandatory human enhancement	Opposition to exclusion of people with disabilities from space missions
Allows the concept of moral uniqueness of the space environment	Rejects the concept of moral uniqueness of the space environment
Criticism of living conditions in space as dangerous for humans in general due to environmental factors	Criticism of living conditions in space as dangerous for humans in general not only due to environmental factors, but especially for specific groups of people due to the dynamics of power structures and mechanisms of oppression

bioethics, often based on principles, is the concept of the veil of ignorance proposed by John Rawls. This thought experiment assumes that we have no knowledge of our social, economic status or possible privileges. Therefore, without knowing what our position is, we should plan to make society as objective and fair as possible. Rawls here uses the liberty principle and the difference principle. The liberty principle means the maximum of freedom for everyone with respect for the freedom of others. The difference principle, on the other hand, means equal opportunities for all with the proviso that if these cannot be guaranteed, norms should be tailored to the most excluded (Rawls 1999). This concept fulfills its role perfectly, as expected from the methodology of nonfeminist bioethics. It has the highest possible degree of impartiality, abstractionism, and generality. Thus, it is a methodological perspective in line with what feminism interprets as a masculine vision of ethics and bioethics, namely abstractionism and a lack of embodiment. However, as feminism points out, decision-making takes place in a social context in which individuals have an available range of options determined by their specificity, position in society, and other external factors (Frazer and Lacey 1993, 54–55). Similarly to philosophy (von Morstein 1988, 147), bioethics is also gender biased, and while bioethics, from its beginnings in the second half of the twentieth century, could be seen as—albeit not explicitly—marginalizing women by viewing them primarily through the lens of female reproductive biology, at least it did not explicitly recognize women as inferior individuals to men, as did philosophy, which explicitly recognized them as serving only the purpose of reproduction (Sreedhar 2017, 145). "The veil of ignorance" thus ignores the most basic and ubiquitous form of oppression and bias, which nevertheless cannot be ignored.

One of the key categories of feminist bioethics methodology is that of intersectionality, which indicates that people are victims of multiple oppressions interacting simultaneously (Tomlinson 2019). This concept, in relation to women, indicates that we should not talk about one category of women, a specific idea of woman. We should be aware of the complex situation of women and the fact that each woman experiences different types of oppression in her own specific way. The concept of intersectionality also indicates that different oppressions combine to produce a specific type of exclusion and marginalization that cannot be reduced to any single type of oppression (Hunt 2017, 131). The concept of intersectionality is an important enrichment of the methodology of nonfeminist bioethics, which usually fails not only to take into account the diversity of the status of individuals in terms of their autonomy and power, but even more to distinguish between different types of oppression or between different groups of women according to their degree of oppression. However, the category of intersectionality arose primarily from the experience of nonwhite women, for whom gender oppression is a product of its interaction with other types of inequality (Mohanty et al. 1991; Zinn and Dill 1994, 3).

Hilde Lindemann criticizes mainstream ethical theories such as social contract theory, utilitarianism, and Kantian ethics. She rejects the vision of the person proposed by these theories—as an individual independent of others, devoid of relationships; self-sufficient, without attention to her interdependence and need for care; as equal to other individuals, possessing power and freedom on a par with others, endowed with a superior intellect. As Lindemann argues, such an understanding of the person is the concept of a completely rational person, who plays a kind of political and social game, makes treatises, and considers abstract moral rules, which she then applies. However, this is a description that characterizes only a small social group at certain points in its life. As Lindemann notes, not only does this picture fail to describe the peculiarities of many men, but it completely fails to reflect the status and situation of women. Consequently, these three main ethical theories do not describe us, because each of us depends on others and needs the help of others. Most people are not in the ideal situation assumed by the theories. Even those ideal men—and perhaps especially them—require care from others. These others are almost exclusively women, who are thus deprived of even the opportunity for the moral and intellectual development assumed by these ideal theories, because they do not have the moral or intellectual conditions necessary for the development described in these theories. These conditions are taken away from them by a sexist culture.

As Lindemann adds, these theories are not only untrue but harmful at the same time, because many people, including the sick, the disabled, and the uneducated, are unable to live up to the ideals of the perfect moral subject. This way of thinking about a person, which excludes women, but also many other groups, including many men, has also perpetuated the separation of the public and private spheres, where these rules and norms do not work (Lindemann 2019, 88–92). These theories express the problems, interests, and challenges of a certain group of men. Therefore, questions and problems specific to women, such as the problem of dependency, for example, are not posed as a starting point. People who do not conform to these theories are "placed" in the private sphere, "outside the law" of morality. These theories do not reflect other perspectives such as dependency, emotion, corporeality, and particularity (Lindemann 2019, 94, 99).

Not surprisingly, understanding the individual, rationality, morality, and society in this way gave rise to bioethics based on abstract principles and rules. Like these normative theories, mainstream bioethics assumes an atomistic conception of the individual/patient, who is autonomous. This model believes that all people analyze a given medical problem in the same way (the concept of people as "interchangeable"). Feminist bioethics reminds us of the crucial importance of relationships, especially the family context for the patient. Just as normative theories were concerned with a small group of privileged men, so, as Lindemann argues, mainstream bioethics serves the interests and needs of privileged and influential people who have power and de facto autonomy. As a counterbalance to this model, feminist bioethics points to the importance of group relations, as well as to the distortions of power at the same time. Lindemann also mentions that mainstream bioethics does not pay attention to the many forms of oppression, does not analyze the oppression of one group against other groups, and does not analyze power structures. Feminist bioethics makes up for this lack by analyzing the power structures and various forms of oppression that exist in healthcare (Lindemann 2019, 128–137).

Narrative and Case-Oriented Mindset

As we have seen, feminist epistemology, which questions the existence of an independent, autonomous, rational individual who would act as an observer and moral judge, thus rejects the domination of the bioethical perspective by abstract, rational principles and rules. Feminist epistemology emphasizes the determinative role played by the social location of the knower, which influences her way of acquiring knowledge (Grasswick 2018). The reference to narrative and case-by-case analysis in feminist methodology thus stems from the belief that the knower is always located "somewhere," in the specific circumstances that define her cognition and knowledge (Code 1993, 39).[7] This is particularly relevant to the cognition and epistemic experience of marginalized groups, who have epistemic privilege due to their unique first-person experience (Bar On 1993).

The feminist understanding of the case method is different from the casuistry familiar in nonfeminist bioethics. In the latter, the starting point is a concrete case, and its analysis is used to determine which of the abstract bioethical principles will be most relevant. In the more radical version of nonfeminist casuistry, it is not that we have a set of ready-made principles and rules, but rather that the analysis of the case serves to co-create principles and rules appropriate to the biomedical situation at hand, which are formalizations of our moral intuitions (Flynn 2022). This second, more radical version of nonfeminist casuistry can only superficially appear similar to the feminist case method. For in reality, any nonfeminist, even the seemingly most radical approach to the case method, does not provide the tools guaranteed to analyze and discern exclusion and oppression due to gender, race, class, and other feminist categories in a given biomedical situation.

Instead, the feminist approach tends to accept the narrative perspective and case analysis. But such case analysis diverges from the case analysis inherent in nonfeminist bioethics, which also assumes the concept of an independent and autonomous observer and moral judge. Instead, Lindemann proposes a narrative methodology based on a network of interactions and dependencies that formulates moral statements expressing group interest. As Lindemann adds, such a narrative is obviously feminist in its opposition to gender- and race-based exclusionary ideologies (Lindemann 2007, 127). Feminist case-by-case analysis examines individual assumptions, hypotheses, and findings in the context of their replication and reinforcement of a system of oppression according to gender, race, and other categories. As Lorraine Code contends, the burden of proving the neutrality of a cognitive procedure lies with the individual claiming the objectivity and neutrality of that procedure (Code 1993, 31).

A Feminist Critique of Social Contract Theory: Structural Oppression

It is worth paying a little more attention to the feminist critique of contractarianism, which is the dominant ethical theory today, especially strong in thinking about society and politics. Feminism's critique of social contract theory is primarily a critique of the lack of trust, an impersonal conception of human relationships. Relationships between people at all levels are governed by rules and contracts. One of alternatives to the contractual understanding of human relationships is the relationship between mother and child, which is intimate, based on trust, unconditional, and devoid of contract (Tong 1993, 52–55). Feminist philosophers criticize the social contract model, instead pointing to the idea of human solidarity, as well as our interdependence, which completely escapes the social contract theory (Ferguson 2009; Sen 2014).

But even if we accept, as do many feminists, that it is difficult to regard the mother-child relationship as paradigmatic and universal, this does not change the fact that contractarianism as understood by Rawls proposing the veil of ignorance makes it impossible to achieve the ideal of justice. Superson questions the fairness of Rawls's conception of the veil of ignorance because of its failure to take gender (as well as any other characteristic according to Rawls's thought experiment) into account. As Superson notes, gender is an important factor due to functioning in an oppressive society (Superson 2009, 40). Having a particular gender changes individuals' moral situation as well as their judgment.

Feminism, in critiquing the concept of individualism assumed by contractarianism, usually points out that contractarianism works well in regulating relationships proper to men in public spaces. It does not, however, reflect the specificity inherent in the centuries-old situation of women, which often continues to this day, namely relations between persons who are close to one another, unequal, dependent, as expressed in the aforementioned mother-child paradigm (which fits here perfectly as an illustration of the specific relations of the private sphere as opposed to the public sphere). Consequently, as Superson notes, contractarianism describes only a part of interhuman relations, namely male relations in the public sphere between strangers (Superson 2009, 40–41).

The category of oppression is one of the key explanatory categories in feminism in general and in feminist bioethics, including the bioethics of space exploration, in particular. What we are interested in in this book, when it comes to oppression, is structural oppression, which is the result of power relations arranged in such a way that they exclude and limit parts of society. Thus, it is not oppression intentionally implemented by individuals against others, but rather the type of social structures in which certain individuals and groups simply partake in the benefits that an unequal and unjust system creates for them. Understood in this way, oppression is an outgrowth of privilege (Silvermint 2017, 50–51). Feminist bioethics views the concept of the social contract as an oppressive concept that assumes abstract relationships between unequal subjects and simply excludes some of them as traditionally marginalized groups. This way of understanding oppression is central to feminism, because it emphasizes that oppression is the result of unjust structures, not individual—even if defined as wrong and unjust—actions.

It is not a problem to point out the existence of privileged groups and groups experiencing oppression if we pay attention to the situation regarding biomedicine on Earth. Typically, men are in a position of privilege over women when it comes to access to healthcare in categories such as the availability of expensive tests and procedures, or to appropriate drugs tested on men, which is related to the marginalization of women in clinical trials, as well as the recognition of certain types of diseases as typically male diseases at the expense of the marginalization of symptoms of the same diseases in women (Ballantyne 2022). Similarly, in countries with problems of structural racism, white people are in a privileged position over nonwhite minorities (Russell 2022). Finally, almost everywhere, people in the upper socioeconomic classes can expect better medical care than members of the lower classes.

Can an analogous oppression be expected with respect to biomedical issues of interest to the feminist bioethics of space exploration? It can be presumed that due to historical oppression and exclusion on Earth, future astronauts of long-term deep-space missions, as well as possible space settlers, will come from privileged groups; that is, white males may predominate, as is currently the case in space flight. Other groups, such as people with disabilities and women, may experience oppression and exclusion on a scale even greater than that known on Earth.

Constancy and Variability in Feminist Methodology

Despite feminism's common concern for the excluded and social justice, feminism fosters a multiplicity of opinions, even conflicting ones, on the same issues. This multiplicity of feminisms may pose a philosophical challenge (Cameron 2020, 8) and raises questions about the relationship of feminisms to concepts such as sustainability and change (Hester 2018). What is changeable and what is permanent? Can it be said that certain biological or, more broadly, natural elements are immutable, while cultural and social elements are changeable? In the methodological perspective proposed in this book, the assumption of permanence of any elements is rejected. The assumption of permanence and immutability can generate the postulate of not changing and not touching what obtains such immutability status. We therefore reject, as does xenofeminism, the idea of the fixity and inviolability of what is biological and natural (Hester 2018, 19–20).

A previous book on the bioethics of space missions (Szocik 2023) devoted considerable attention to criticizing the concept of human nature. Feminism is usually also opposed to the concept of nature understood as something immutable. It is this belief that has been used to justify the inequality of women vis-à-vis men, and the concept of naturalness and nature is also used to discriminate against and deny the same rights to homosexual and gender nonbinary people.

Feminist bioethics, as we argue in chapter 5, is also often skeptical of the idea of human enhancement, but not on the basis of the idea of the immutability of nature—for that is what it usually rejects—but because of the perceived and predicted inequalities and social injustices associated with unjust power structures (however, this is not the only one of the objections raised against the idea of human enhancement by feminists).

This issue of variability and constancy is often presented as a distinction between nature and nurture or nature and culture (Watson 2017, 71–72). This distinction is important for bioethical methodology, since the status accorded to what is natural, as well as what is considered natural, can be used to justify the right to implement or prohibit the implementation of particular biomedical procedures. Examples include abortion, euthanasia, and assisted suicide, but also assisted reproductive technologies and human enhancement. The bioconservative paradigm tends to equate the natural with the fundamental and unchangeable. Consequently, if something is considered natural, it should not be modified. This way of reasoning has a number of weaknesses, the main ones being the treating of diseases that "naturally" occur in the population, and the occurrence of evolution through natural selection and mutation, which cause constant changes in what bioconservatives consider natural. Despite these weaknesses in the argument, the position is used to justify bans on abortion and euthanasia, as well as human enhancement.

Another problem, which Helen Hester also rightly points out, is the risk of equating certain inclinations and actions with a particular gender. Hester cites the context of ecofeminism, where what is feminine is associated with protection and care for the environment, while what is masculine is associated with its destruction (Hester 2018, 39). Such a gendered way of thinking has certain advantages, as it makes it possible to demystify and deconstruct certain mechanisms and structures in the patriarchal world. However, in the next stages of the analysis, attention should be given to the problem and the concept of improving unjust relations, which in turn would require a nongendered way of thinking. Gendered thinking was also evident in science, especially evolutionary science and psychology, where women and even non-human female animals were seen as passive, attributing activity to men and males (Carter and Fisher 2017, 803–804).

Women in Space: Problems for Health

Almost already a classic theme on gender addressed in space policy is the belief that a female spacecraft crew is attractive due to the statistically smaller size and weight of women than men, which is translated into lower requirements for the necessary resources and, consequently, a lower spacecraft mass. Less clear is the expectation regarding morality and behavior, which nevertheless refers to both the stereotypical and statistically confirmed lower level of aggressiveness and violence committed by women compared to men.[8] Geoffrey A. Landis proposes an all-female crew for future space missions to Mars because of these two parameters (Landis 2000).

Despite this, women are still a minority in space missions. Indications are that women are more vulnerable to the negative consequences of losing bone and muscle mass in space due to altered gravity. Women are also more likely to suffer the negative effects of cardiovascular deconditioning, as well as experiencing motion sickness more. Women are also more susceptible to cancers caused by cosmic radiation, especially lung, breast, and ovarian cancer. Women's greater sensitivity to the negative effects of radiation in space makes the total period of time allowed in space for women shorter than that for men (Prysyazhnyuk and McGregor 2022, 130–131). However, this more susceptible physiology of women to the negative consequences of space radiation should not be an argument for excluding them or limiting their presence during long-term exploration missions, but a rationale for increasing bio-medical and technical protection of women from the negative effects of radi-ation. The concept of human enhancement discussed in this book is one such countermeasure.

Many studies on women in space could help improve healthcare for women on Earth, including for cancer, osteoporosis, and aging. The latter is impor-tant for women, since most of the population in general, and older people in particular, are women, who statistically live longer than men. Women, how-ever, are usually excluded from research in space, in line with the situation in clinical trials conducted on Earth (Shayler and Moule 2005, 343–356).

Women's representation in space missions should be significantly increased, not only for the sake of gaining more knowledge about protecting women's health on Earth, but also with an eye toward future long-term deep-space missions. This is because there is a risk that the knowledge gained from past and current orbital missions due to the dominance of male astronauts will be useful primarily for men, and that knowledge about women's health on long-term space missions will be limited and inadequate, as is currently the case for many areas of healthcare on Earth. An increase in the intensity and scope of women's clinical research will indicate a serious plan to include them in future space activities, which are currently reserved almost exclusively for men. Nor will their underrepresentation in clinical trials be able to be taken as an argument for excluding them or hindering their participation in future space missions when they become more accessible and, for various reasons, attractive.

On the other hand, it is worth noting that the space environment is very dangerous for everyone, not just women. The dangerous factors character-izing the space environment and their translation into bioethical consider-ations that take into account concepts even as radical as human enhancement, including gene editing, have been discussed elsewhere (Szocik 2023). Mark J. Shelhamer and Graham B. I. Scott list a number of risks to human health in space, such as the harmful effects of cosmic radiation, the danger of inhaling contaminated air, decompression sickness, behavioral problems, medica-tion toxicity, inadequate nutrition, sleep problems, changes in vision, effects of dust, cardiac changes, and changes in bone density (Shelhamer and Scott 2021, 800). The biggest challenge to human health and life will be a mission to Mars, which is among the highest-end space exploration missions. It will take a very long time to return to Earth in emergency situations, and at cer-tain stages of the mission, such as during interplanetary trajectories, it will not be possible at all. Obtaining new medical supplies will be impossible, as will receiving any support from Earth (Shelhamer and Scott 2021, 803). The space environment is extreme and negatively affects and threatens all physiological, psychological, and behavioral human systems (Ball and Evans 2001; Prysyazhnyuk and McGregor 2022). In contrast, the degree of negative impact of the space environment depends on the scenario under which the space mission is planned.

Significant differences in the extent of harm are indicated by a comparison of the accelerated Mars mission with the standard Mars mission. The accelerated mission, in which the stay on the surface of Mars would be 30 days, guarantees a 2.9 times lower risk of loss of crew life, a 4.7 times lower risk of a serious medical condition, a 19 percent better Crew Health Index, and a 1.5–2 times lower risk of cancer caused by exposure to cosmic radiation than in the standard mission, in which the stay on the surface of Mars is 560 days (Antonsen and Van Baalen 2021). Such a significant shortening of the stay on Mars may be possible for a reconnaissance mission. But for a mission oriented toward longer exploration and exploitation, in terms of a future permanent stay in space, an accelerated mission scenario is unrealistic.

And this is where feminism comes in. It is feminist philosophy that makes it possible to ask whether, given the health challenges, such missions are worthwhile at all, what their actual intended purpose and effects on humanity are—if any—and whether, if such a mission should nevertheless be carried out, it should not be an accelerated mission that, for the sake of human health, ignores the potential political, military, or commercial benefits offered by a standard mission.

Notes

1. Just as bioethics in general cannot ignore the problem of power structures by means of, among other things, principles of justice (Marway and Widdows 2015, 170), bioethics concerning space exploration should do the same.
2. Henk A. M. J. ten Have refers to mainstream, nonfeminist bioethics as "bizarre bioethics," i.e., bioethics that focuses on bizarre cases that are rare and unrepresentative of the problems afflicting all of humanity in healthcare (ten Have 2022).
3. On abortion in space, see also (Milligan 2015, 16–17; 2016).
4. See also the problem of symbolic racism in the United States (Anderson 2022).
5. A similar position is presented by Johnson-Schwartz, who argues that the political philosophy applied to human habitats in space should be very demanding and should not relax moral norms due to the demanding conditions of the space environment (Schwartz 2022).
6. See *A Cyborg Manifesto*, in which Donna J. Haraway (1991) suggests not judging science and technology negatively, but accepting them as new forces shaping our reality and using them. Haraway also adds that the boundaries between culture and nature are not rigid, but are constantly modified, shaped, constructed, and thus there is no such thing as an unchanging,

pure nature. Many people are, in a sense, "cyborgs" through the use of more or less body-related technologies. Certainly this idea is useful in the context of the discussion of the idea of human enhancement, but it is worth noting that it is widely assumed by many bioethicists and nonfeminist ethicists advocating human enhancement. It is therefore not an exclusively or typically feminist idea.

7. According to feminist epistemology, knowledge is always situated (Deitch 2021).
8. See also Gentry and Sjoberg 2015 on dominant narratives on violence committed by women.

3

Power and Exclusion in Space Exploration

The Problem of Power

Introduction: The Pervasive Power of Men over Women

Critiquing power structures is one of the traditional themes of feminist philosophy and bioethics. The problem of power is the problem of the establishment (Marway and Widdows 2015). The issue of power is of fundamental significance to other issues such as exclusion, discrimination, and exploitation. The ability to exclude others, to oppress, to exploit, and to subjugate requires power, however that power is understood and whatever level and degree it implies. Without power structures, there are no other categories. In order to subjugate women under sexism and patriarchy, the lower classes to the upper classes, and races considered inferior to races considered superior, the parties carrying out such subjugation must have power.

Power structures favoring men have consequences in all areas of life, including healthcare and biomedical issues. Healthcare contains gendered structures. The underrepresentation of women in the broader space industry can only exacerbate the genderization of medical care, in this case space medicine. This situation leads to a kind of feedback loop, where insufficient study of the physiological effects of space on women's bodies can contribute to the underrepresentation of women in space. With respect to nonspace medical care, the preference for men, as well as the preference for, and dominance of, the male point of view—for example in bioethical discussions and decisions or masculinized abstract normative systems—leads to effects such as decreased access to abortion, which makes the already challenging situation of a woman seeking an abortion more difficult, and the de facto exclusion of women from many clinical trials (Lindemann 2019), including, but not limited to, pain research, cardiovascular disease, and Parkinson's disease research using biased artificial intelligence, as well as underrepresentation and inadequate knowledge of treatment for pregnant women (Ballantyne 2022). Women's

Feminist Bioethics in Space. Konrad Szocik, Oxford University Press. © Oxford University Press 2024.
DOI: 10.1093/9780197691076.003.0004

experience, pain, and illnesses are marginalized and interpreted in stereotypical ways, leading to male-centralized medicine with negative consequences for women's health (McGregor 2020).[1]

It is worth adding that the ever-recurring topic of abortion around the world, where, in the case of the United States, the Supreme Court reversed *Roe v. Wade*, can also be read in terms of categories such as rights and property typical of the male contractual approach to reality. According to this contractual approach, human life must be governed by abstract rules. The problem, however, is that almost the only actors who set the terms of contracts are usually white men with power. They extend their power also, and perhaps especially, to women's biology, especially reproductive biology. As Rawlinson argues, the female body has always been a battleground between men for power and ownership (2016, 107).

Another example of men's power over women in healthcare is the medicalization of birth.[2] Whereas just over one hundred years ago in the United States and other Western countries only a small proportion of births took place in hospitals, nearly all now do. In the process, midwives have been systematically displaced by male physicians (Stone 2007, 168–169). The appropriation of childbirth—recognized as a pathological phenomenon, as a kind of disease and risk—by men, who played the role of experts, coordinators, and managers of childbirth, made possible the transition to birth control in terms of decisions on the acceptability of abortion, in vitro fertilization, and the use of reproductive technologies. This appropriation built on previous knowledge power structures in which the only people who can acquire knowledge, but also the only moral (rational) subjects, are men. From the perspective of feminist bioethics, biomedical technologies, including reproductive technologies, are not bad in themselves. Instead, technologies are not neutral and always serve someone and are under someone's control. In the case of reproductive technologies, they are controlled by men, while their main addressees are women (Rowland 1989, 356; Paton 2022; Scott 2022; Turkmendag 2022). Feminist critiques of reproductive technologies point out that it is women who bear the greatest burden of testing, introducing, and using reproductive technologies, which are additionally presented in terms of therapy, prevention, and a way to improve the quality of women's lives (Corea 1988, 134). The intersectional perspective shows how reproductive technologies can replicate and reinforce the dominance of already privileged groups, replicating the racist paradigm of domination by white middle-class people who instrumentally exploit the bodies of nonwhite women, this time using biomedical technologies (Harrison 2016).

Men's power over women is also evident in practices regarding women's style and dress. According to Rawlinson, women's dress practices in Saudi Arabia, as well as restrictions on their travel, are designed to make women invisible and immobile in public spaces. However, Rawlinson adds, the role played by power structures, specifically biopower, in Western democratic countries should not be marginalized. In the West, the female body is both subjugated and restricted, but at the same time made an object of profit in capitalism in the clothing and fashion industries, among others.[3] In contrast, restrictions analogous to those in Saudi Arabia involve impeding mobility through the promotion of high-heeled shoes, which Rawlinson argues reinforce the image of women as not being good candidates for positions in business, politics, or science (2016, 64–65).

The problem of power is not only about sexism but also about characteristics other than gender, which have been considered lower elements of dichotomous thinking in Western patriarchal philosophy and culture. These other elements, no less important and no less severe for the excluded and marginalized, are race and class. A valid critique of first- and second-wave feminism focused primarily on sexism makes it imperative that the analysis of oppression be equally concerned with all the excluded regardless of the characteristic that serves as the basis for exclusion. Focusing only on sexism to the exclusion of racism and classism would lead to the risk of an exclusivist concern only with the status of white middle-class women (Griscom 1987, 88–89, 93).

Power is associated with the possession of privilege. Such privileges are present in all areas of life, including in bioethics, where they are distributed by gender, race, and class. In terms of gender, it is men who tend to be privileged, while in terms of race, it is white people, and in terms of class, the wealthy. As Alison Bailey notes, privilege is unearned; the privileged group is usually unaware of its privileged position relative to others (2021, 1). The privileged group, consciously or not, imposes its narrative and takes it for granted. This is the case in current space policy, which is dominated by a small, privileged, masculinized, and Western group of space exploitation advocates (Billings 2023, 26; Nesvold 2023).

There is a risk that the long-term human presence in space will replicate the structures of power and discrimination familiar from Earth. The peculiarities of the space mission environment, how space policy is practiced, allow us to assume that unequal power structures between the genders and races will be intensified in space.

The categories of race, class, gender, and ideas about the environment discussed in this chapter are intersectionally intertwined with a critique of those colonial and imperial narratives that underlie space exploration and that contribute to exclusion.

Spatial and Material Dimensions of Power

In addition to sexism and racism, which are the foundations of structural injustice because gender and race are often the criteria for the division of labor (McLaren 2019, 167),[4] the problem of power has a specific spatial and material dimension. This is an element that is likely to be particularly important in the context of space exploitation and exploration. Any human activity in space will require a guaranteed supply chain from Earth, regular transportation between Earth and the space object. Transportation will be of strategic importance as long as the human presence in a given space object does not achieve self-sustaining status. Human activity in space will therefore depend on the supply of all resources, including energy sources, from Earth. Because of the strategic importance of these supplies, control of the supply chain will be related to the accumulation of power and its deployment. Since control of the supply of strategic commodities, especially energy sources, is arguably the greatest form of power in the world and enables direct and indirect influence on global politics, an analogous situation can be expected in space, where the number of alternatives appears to be lower than on Earth.

Amitav Ghosh's analysis of power dynamics on Earth, currently centered around control of the global energy supply chain, may provide a basis for such predictions of future power dynamics in space. As Ghosh asserts, the main reason for blocking the transition from fossil fuel-based energy production to renewable energy production is not so much neoliberal capitalism as the ability to control where material energy is produced and transported. According to Ghosh, the ocean transportation of oil controlled by the US Navy plays a key role. Unlike oil, renewable energy cannot be materialized and thus controlled (Ghosh 2021, 108–109).

Depending on the stage of development and the goals of space exploitation and exploration, the power dynamics centered around the supply chain from Earth will have specificities appropriate to the type of mission. In the case of commercial missions by a single country, entities that monopolize transportation between Earth and a given object in space may gain a negotiating advantage. If they are private entities, they may induce the government to grant them privileges and priority access to resources. In the situation of the more distant future, when space exploration acquires the form of a permanent base in space, the dependence of this base, a potential space settlement, on Earth can be used to control the way it functions, its political organization, or, finally, it can inhibit the possibility of its independence from Earth by blocking the development of energy sources in the space base that guarantee independence.

Family, Power, and Feminism

The spatial and material dimensions of power outlined above describe a global structure that is sure to be replicated in space exploration. But the starting point for this global power structure is the most basic unit of organization of social life in Western cultures, the family. The nuclear family is where the oppression of women begins and takes place (Okin 1989). But the family is also responsible for many other negative phenomena. As Alison Phipps argues, the nuclear family is the foundation of capitalist patriarchy. This is because the sexist model of the nuclear family guarantees free care for children, as well as often the elderly and sick, who under capitalism do not have to be paid, and thus enables the exploitation of workers who can devote themselves to work while another member of the family, usually a woman, takes care of the upbringing and care. It also generates racism and classism, ascribing superiority to the Western, white family model over the social life organizations of colonized peoples (Phipps 2020, 8). Following Marxist and socialist feminism, we can add that the nuclear family is the cause of the exploitation of both people and all resources and the environment. It is also a source of tension and conflict because of the role played by the concept of ownership, inheritance, and the idea of leaving as many resources as possible to one's biological offspring, among other things (Federici 2021). The nuclear family, however, is crucial to capitalism because of its ability to put men to work and keep women as caregivers, who are undervalued and underpaid but strategically vital to maintaining the capitalist social model (Scott 2018, 84–85).

The family is also the main site of women's oppression, which determines their future fate and teaches male-subordinated social roles. The family is a private sphere, set aside for women and children, which in a patriarchal society separates women and children from the public sphere of men. It is worth noting the forms of discrimination against women in the evolution of culture, especially in the context of the evolution of cooperation, which in practice was the domain of men, assigning a marginal role to women.[5] It is worth noting the rather traditional explanation of the evolution of cooperation and morality, which is an example of an ideal theory that pays no attention to the actual power structures that discriminate against women. Léo Fitouchi and colleagues offer such an idealistic vision of the evolution of cooperation (Fitouchi et al. 2023).[6] An alternative to this idealistic model is a nonidealistic feminist critique focused on the puritanical regulation of sexual behavior and those elements that particularly concern women (Appel 2015). Taking the feminist critique as a starting point, the theory of Fitouchi and colleagues shows that the evolution of cooperation in the model presented by the authors is the evolution of cooperation between men, not including women, treating women instrumentally, which corresponds to the patriarchal nature of the mechanisms described in this theory.

The authors cite the example of norms that mandate the covering of the body by women caused by concern for the possible loss of self-control by men, which can consequently undermine cooperation in society. The authors even propose analyzing the prohibition of premarital sex in terms of proximate behavior intended to strengthen cooperation. Even if this was the real adaptive value of the prohibition of premarital sex, the burden and consequences were on the woman, who ceased to be a virgin and possibly became pregnant. Women usually did not have the right to make decisions about marital matters, telling examples of which are the practices of bride abduction and arranged marriages found in various cultures (Vandermassen 2008).

The cooperative component of these and similar practices is insignificant compared to the sexist desire of men to dominate women and exclude them from public life. Thus, if the restrictions discussed by the authors that characterize puritanical morality did indeed reinforce cooperation, it was a cooperation between men and men. Women were excluded because cooperation is a feature of public space, and the place of women in sexist societies was the private sphere, where cooperation occurred spontaneously, based on kinship ties. Moreover, women were usually subordinated and dependent on men controlling resources, so they were not subjects of the evolution of cooperation (Vandermassen 2005, 187).

According to the sexual selection model proposed by Patricia Adair Gowaty, males sought to control women, while females sought to repel that control (Gowaty 1992). Thus, the model proposed by Fitouchi and colleagues to explain cooperation is a model that excludes women, as well as the populations conquered by the colonizers. Gowaty adds, moreover, that the underlying assumption in Darwin's theory and cultivated to this day about the dominant role played by male-male competition is, if not wrong, overemphasized, at the expense of undervaluing the actual important role played by female-female competition. Sarah Blaffer Hrdy (1999) argues that the evolution of women in the context of traits such as dominance, rivalry, and competition has proceeded in part differently from the evolution of other primates, which only confirms that women are the most oppressed of the female primates.

It is worth remembering that the practice of covering women, usually coupled with the prohibition of their movement without the company of a male guardian, leads to women in these cultures becoming invisible and immobile—in a sense, ceasing to exist (Rawlinson 2016). Thus handicapped, they become easily controllable and cease to be competitors (Gowaty 1992). Deprived of any place in public space, women are forced to take care of the home, relieving men of these responsibilities.[7]

Also worth bearing in mind is the feminist critique of the theory of biological and cultural evolution. Feminist social epistemologists and philosophers of science remind us that social and cultural factors shape the way scientists think and do science. This was also true of Charles Darwin and the stereotypes he reproduced about the role of gender, evident in his theory of sexual selection (Nelson 2017). This applies not only to scientific theories but also to religious and ethical systems, including the concept of puritanical morality. While many of the mechanisms described by Fitouchi and colleagues can be explained in terms of the evolution of cooperation, there is a strong rationale for the hypothesis that explains the mechanisms regulating women's behavior and practices in terms of their exploitation and domination by men. The practice of covering women is an objectification of the woman, who for the man was never an equal to him—such status was only held by another man. Therefore, the evolution of cooperation and its effect, the social contract, is really a sexual contract between men (and de facto also a racial contract, if we take into account the exclusion and colonization by white Europeans of the rest of the world).

In conclusion, the missing element in the theory of cooperative evolution proposed by Fitouchi and colleagues, as well as in other mainstream theories that explain the evolution of cooperation and morality without reference to power structures, is the omission of this component of exploitation of women and their objectification. But even if these practices were to actually enhance cooperation, the entire burden falls on women, not men, who are simultaneously stigmatized for distracting men from publicly relevant issues. The evolution of cooperation that has taken place in this way requires an explanation of why evolution has discriminated against, marginalized, and placed a burden on women. While the mechanisms in question may be adaptive in an ideal society, in a nonideal—patriarchal—society they are a tool of oppression and control, adaptive only for a select group of men. In the abstract world of evolutionary theory, females invest more in parental care, but this biological asymmetry of the world of biology in a nonideal society has become a justification for the cultural and social asymmetry between men and women (Vandermassen 2005, 78–79). Interestingly, the social naturalness of this asymmetry was assumed by religious systems, which can be interpreted as supporting the mechanisms favored by sexual selection. It is worth adding, however, that reproductive morals are a better indicator of religiosity than co-operative ones (Van Slyke and Szocik 2020), which seems to minimize the cooperative value of religiously sanctioned restrictions, especially affecting women. It is difficult to see the gender socialization manifested in restrictions on women's freedom and choice as having any relevance to the evolution of cooperation other than sexist exploitation and subjugation by men. This is es-pecially true of regulations, including penalties on sexual behavior and repro-duction, which were almost exclusively imposed on women (Vandermassen 2005, 149–150). If these regulations were meant to promote cooperation, why has male sexual behavior not been equally regulated throughout history? Both historically and often even today, it is the woman's body, not the man's, that has been, and continues to be, controlled and is of particular interest to both various ideologies and the state (Ralston 2021).

It is also worth remembering that the model of cooperation, at least with re-spect to the Western world, was based on the model of the citizen as a person with qualities such as self-control, which are supposed to indicate a willing-ness to cooperate responsibly. However, these qualities describe a man, not a woman, who has traditionally been associated with carnality and emotion-ality. Moreover, these traits, insofar as they were attributed to a woman, usually represented her disadvantages, not her advantages (Holland 1990). Thus, as can be seen, the concept of the nuclear family has far-reaching consequences, as it guarantees the maintenance of the traditional subordinate role of women, while at the same time it is meant to protect men from society's interference in private affairs. The more patriarchy and sexism in society, the more criti-cism and prohibition of alternative forms of social organization to the nuclear family, such as civil unions, or marriages and partnerships between sexually nonbinary people, or the adoption of children by LGTBQAI+ people.

Gender in Space Exploration

The problems of power and gender on Earth outlined above are evident in the politics and philosophy of space exploration. Space missions are asso-ciated with masculinity and patriotism. The latter is often permeated with militarism. In many films about space missions, we typically see a male pro-tagonist portrayed as the hero, although the lack of women in cinema in many key roles is not just the domain of space films. Many areas of life are heavily masculinized, but the area of space exploration seems to be one of the leading ones.

Is it legitimate to assume that space offers new opportunities for humanity in terms of the possibility of creating a society based on principles of justice and equality? Is it reasonable to conclude that humanity in space will have the opportunity to create a society that treats men and women equally? When it comes to politics on Earth, equality policies, affirmative action, and quotas for women, among others, are always secondary and reactive. Even in those societies where the degree of equality is greatest, women have been discriminated against and equality—not necessarily always complete in all areas—has only been artificially introduced. In the case of space missions, humanity, being equipped with knowledge of equality and the history of discrimination and oppression, as well as familiar with all waves of feminist thought, has an opportunity to remodel and reevaluate its thinking about sex and gender. But space does not necessarily offer more opportunities than Earth, and changing the environment alone does not imply pursuing policies that respect justice and equality—especially in the current situation, in which inequality between men and women is quite entrenched in the cultures that dominate modern space exploration.

This opportunity applies to all traditionally marginalized and excluded groups. Among these many groups, not just women, it seems that space exploration offers a special opportunity for yet another group, the disabled. As I argue in chapter 4 on disability in space, the space mission environment has the potential to remodel our thinking about disability—this despite a certain paradox, which is the existence of very restrictive requirements for space mission candidates regarding fitness and physical ability.

A reference to the debate between equality and difference in feminism may be useful here. However, it is worth remembering that this is only one model for organizing public life. Feminism grew out of the desire to give women equal rights with men. But attention was soon drawn to the fact that, in essence, this ideal of equality consists of a desire to identify with a pattern that is masculine, produced throughout history by men, and that expresses a male point of view and a male interest. The ideal of women's equality in its minimalist version is to grant women equal rights and opportunities in a culture shaped in a masculine style. Women simply inherit this culture and are forced to function within a masculine system of hierarchies and symbols. Therefore, an alternative to the feminist paradigm of equality is the paradigm of difference, which, assuming an underlying equality of opportunities and rights, would open up a new symbolic and axiological space for women ("positive female identity"), who would not have to adapt to male culture (Stone 2007, 133–135).[8]

Both options are problematic for feminism. The liberal option is problematic because it means allowing women to participate more or less equally in public life organized, arranged, and controlled by men. Consequently, women usually have to adapt to the established rules and customs. At least on the surface, the radical option, which emphasizes women's difference, should seem more attractive to feminism. However, this perspective implies the existence of some traits unique only to women or otherwise associated with women, which in principle becomes difficult to ontologically separate from the idea of femininity and carries the risk of essentialism.

The history of space missions to date, however, shows that the opportunity suggested above to design a new reality based on equality may be as difficult, and perhaps even more so, than on Earth. The attitude of NASA and politicians, as well as the organizers of US space missions during the space race with the USSR, discriminated against women incomparably more than the USSR. Prior to the moon landing, US actions were reactive to the USSR's pioneering achievements in space. The United States reacted to the sending of the Sputnik satellite as well as to the sending of cosmonaut Yuri Gagarin. In contrast, they did not react to the USSR's sending of the first female cosmonaut, Valentina Tereshkova, into space in 1963. Margaret A. Weitekamp suggests that among the many possible causes was a sexist and patriarchal understanding of gender (however, Weitekamp does not use these terms). She argues that women in the United States were seen as a delicate gender in need of protection, and the possible death of a woman in space could lead to the cultural destruction of that image.

Sending an American woman into space was never seriously considered until 1983. It was prevented by a macho culture dominated by test pilots and a military focus. A woman's place was in the home and with family, not public space, especially one as demanding and militarized as the space industry. Since sending Tereshkova into space demonstrated that operating Soviet space vehicles must be easy (which at the time struck at the stereotype both of men's superior technological and engineering abilities and of the space industry being reserved for men only), the fear in the United States was that analogously sending a woman into space would suggest to the world that the complexity and technological sophistication of the United States was so low that it could be handled by a woman. Not sending women into space may have preserved the myth that the difficulty and technological sophistication of the United States could only be handled by men. Also important was the belief that the appropriate sphere of a woman's activity was the private, domestic sphere, which is why Soviet women were portrayed in the United States as masculinized and breaking this American, sexist stereotype (Weitekamp 2004, 2–4).

Two points are worth noting here. The first is the historical context. The 1950s and 1960s preceded the second wave of feminism. It was a period when women were deprived of many of the opportunities they have today. But the second point undermines this attempt to explain the sexism of the time with the context of the era. The USSR was less sexist for various reasons, and this was manifested by the involvement of women in professions considered masculine in various spheres of activity, not just the space industry.

NASA's masculinist culture was merely a reflection of the sexism inherent in all American (and not just American) culture. The selection as test pilots of not just men, but specific men—brave, brash, well trained, disciplined, and order-obeying—narrowed the candidate pool to military fighter pilots (Weitekamp 2004, 42–43).

In a sense, this decision was the result of structure rather than an arbitrary decision by the mission planners and organizers. Feminist philosophy exposes the importance of structures in sexism, racism, and classism by showing that oppression is deeply rooted in politics and culture, but also mentality. As Bailey shows, the privileged are usually oblivious to structures of oppression based on unearned privilege. A privileged position is often assumed to be natural and taken for granted (Bailey 2021).

The problem of sexism in NASA culture was reinforced by the danger and risk that characterized test flights and later space flight. In this respect, the aerospace environment differs from other masculinized environments, such as academia and politics. The colloquial notion of men as risk-prone and women as incomparably more cautious than men is supported by the biological explanation of female caution in terms of "staying alive" theory (Benenson et al. 2022). Even if this theory is true, women's tendency to be cautious and risk-averse is in large part driven by the culturally and socially centuries-old division of labor between women culturally predestined to care for children and men culturally segregated from that work. This tendency to explain certain supposedly feminine preferences in terms of biology and psychology shows the problem of philosophy of science (Birke 1986). Science too is socially constructed and also around the concept of sex. An example of such a construction is the concept of studying humans by sex rather than simply as human beings (Szocik 2022). However, in order to study humans always or usually as female or male human beings, it is necessary to have relatively precise and comparatively unambiguous characteristics of both sexes. This is where sexism and a number of the numerous biases in science already begin.

Thus, even though women may possess a set of predispositions that they themselves may believe make them less than ideal candidates for the jobs of test pilots and astronauts than men, such a situation does not illustrate a presumed female psyche so much as reflect a sexist structure in society and culture. A scientific explanation of certain regularities in sex terms runs the risk of biologizing social behavior that is the result of socialization and social structures, not the result of biology.

However, if we are already operating within a sexist structure that biologizes many social and cultural elements and tends to emphasize male dominance due to their presumed biology (for example, greater physical strength and a tendency toward aggression), it is worth noting that female psychology and biology may offer some potential to give women an advantage in competition with men in the context of space exploration, given the variation in characteristics like body size and physical resistance within groups of different sexes. These are advantages that cannot be ignored, especially in such a demanding environment as space flight. Many psychological tests in the late 1950s and early 1960s showed that women were better suited to space missions in terms of their resilience to being in isolation and cramped spaces. Among other things, the tests showed women's greater resilience to sensory deprivation, as well as the absence of hallucinations when isolated for several days, in contrast to the men involved in the experiment. Women also tend to be smaller and weigh less than men. This translates into lower energy requirements, and the lower weight of both astronauts and cargo also means less weight for the entire spacecraft and less propellant needed (Weitekamp 2004, 64–65).

But despite these advantages, the possible presence of a woman in a spacecraft was also seen as a challenge and source of potential difficulty. This included menstruation, hence the emerging ideas aimed at even eliminating it. According to Johnnie R. Betson and Robert R. Secrest, female menstruation not only excludes women from the role of astronauts but also prevents studying them fully because of the physiological instability of women as research subjects caused by menstruation. The authors also point to the negative effects of menstruation on peripheral vision and coordination. They also highlight the positive correlation between menstruation and the risk of mental illness or suicidal tendencies (Betson and Secrest 1964).

Another problem was female sexuality and the fear of the emergence of both sexual relations and other tensions between women and men—hence one of the ideas that the female astronaut should be the wife of the current male astronaut (Weitekamp 2004, 68). These accounts described, among others, in *Look* magazine published in 1960 on the potential role of women in NASA's space program resemble the characteristics of female myths in a patriarchal culture described by Beauvoir in 1949. Beauvoir described the myths about women that are the foundation of Western culture, including women as nature and as mystery. She also devoted considerable attention to female menstruation and its perception in a male-dominated culture (Beauvoir 2011).

As we know today from a historical perspective, the US political authorities and NASA authorities did not approve of a woman's participation in space missions during the Cold War even in a manner analogous to Tereshkova's mission.[9] Jerrie Cobb's efforts to become the first female astronaut and to include women in the space program failed. NASA emphasized the need to meet two requirements: experience as a jet test pilot and an engineering degree (Weitekamp 2004, 160). The first requirement in particular excluded women because it was aimed at military pilots, which women could not be. Regardless of this official explanation, however, there is much to suggest a sexist motivation and problems with anti-feminist perceptions of gendering roles, as suggested by numerous statements by, among others, John Glenn, who spoke repeatedly, including at the hearings of the House Committee on Science and Astronautics in July 1962, about the existing social structure and traditional gender roles in which it is men, not women, who are destined to participate in wars, fly airplanes, and then design and test them (Weitekamp 2004, 151).

In conclusion, the status of women in the space mission environment shows how sexism can operate on many levels, from individual mentality to the structural exclusion of women. In this environment, women are a problem, not an opportunity, and even if their participation is no longer formally prohibited as in the past, women participated in only 136 space flights in the U.S. (men in 772), 8 flights in Russia (men in 275), and 4 in China (men in 28) (Astronaut/Cosmonaut Statistics 2024). As we have shown elsewhere, affirmative action and increasing the percentage of marginalized groups, especially by gender and race, in the power structures does not remove multilevel sexism and racism (Szocik 2024).

Exclusion in Space Exploration and Exploitation

Exclusion in the context of space exploration and exploitation means many processes and mechanisms. It means both the exclusion of women on many levels from this traditionally male environment, and the participation of only the richest countries in the new space race. The richest spacefaring countries, through commercialization of space, not only gain a new source of wealth, but perhaps a solution to social or environmental problems. The neediest countries of the Global South are excluded from this new area of human activity.[10]

The problem of exclusion in the space mission context has some parallels to global exclusion in the terrestrial context. Its roots go back to colonialism. A feminist analysis of colonialism reveals how Western colonists affected not only natural resources but also the sense of identity of the colonized. In the case of Africa, they introduced the category of "black" as the supreme category, which was considered the most important, constituting the essence of a person, instead of remaining something irrelevant and secondary, which is basically what skin color is. This category was then associated with something inferior and subordinate to the category of "white," whereas African people did not perceive themselves in the precolonial period through the lens of skin color (Tong 1997, 83). In recent years, nonwhite scholars of racism in particular have highlighted the pervasiveness of racist patterns in virtually all spheres of social and cultural life, from science and technology to police profiling (Roberts 2011; Bonilla-Silva 2022).

The approach to the problem of exclusion varies according to the type of feminist philosophy. Liberal feminism, for example, to some extent repeats the nonfeminist division between classical liberalism and welfare liberalism. While the former emphasizes the importance of equal rights, the latter recognizes that legal equality alone is not enough because different individuals have different degrees of advantage or disadvantage in competing for limited market resources. Therefore, state assistance is needed for the more excluded. Often, formally equal rights for both genders fall short of actual equality in access to various goods because being a biological woman is often associated with being a cultural woman. And the latter is often associated with the private sphere, which in turn is presented in opposition to the public sphere. This is why, as some feminists rightly point out, they postulate the ideal of androgyny, according to which both feminine and masculine traits are valued equally (Tong 2009, 35–36). This would be an alternative to a model in which feminine traits are judged negatively and consequently lead to the exclusion of women.

One of the fundamental consequences of this unequal treatment of women due to biological differences is their secondary position in the labor market. Women earn less than men, work for free at home and in childrearing, and represent a huge proportion of labor migrants, again exploited. The unpaid work of housewives in particular represents an enormous surplus value for global capitalism (Donovan 2012, 203). The stereotype of women as destined or predisposed to care for others, and thus predisposed—at least more than men—to childcare and housework, is maintained. As Phipps argues, these tasks, considered feminine in patriarchy, are treated in terms of love rather than work requiring remuneration (Phipps 2020, 8).

The Multifaceted Nature of Exclusion

The problem of exclusion in space exploration and exploitation is not just about excluding women. There are several levels of exclusion here that are worth noting. The first is the exclusion of women in general. The second is the exclusion of a certain type of women. The third is the exclusion of groups other than women because of their characteristics other than gender (although according to the concept of intersectionality, having a particular biological sex, usually female, reinforces or overlaps with exclusion caused, for example, by a particular social class, occupation, religion, or ethnicity).

Exclusion of Women

The scant representation of women among space mission participants should raise serious concerns for the future. This phenomenon is structural and multifaceted and is not at all about the sheer number of female astronauts. But it also concerns the very thinking about space policy dominated by hegemonic masculinity and thinking about space in terms of national security (Whitman Cobb 2024). To some extent, this phenomenon is correlated with fewer women than men studying STEM. Consequently, the fewer the women who study STEM and the fewer who are interested in issues at the interface of industry, resource exploitation, aeronautics, and military science, the fewer the women who will ultimately participate in space missions. This also translates into less support from women for space exploration (Whitman Cobb 2022). But this lower representation of women has a partial cause in the discrimination and prejudices against women who pursue "typically male" majors, such as the aforementioned STEM.[11] The "stereotype threat" may explain why women do not choose those fields of activity (studies, professions) that are associated with a masculine approach, preferring a more female-friendly social environment (Mason 2022, 38–39). Oppressive and gender-balanced upbringing methods teach girls to assume "girly" and then feminine roles and, by analogy, prepare boys for "masculine" roles, and these patriarchal upbringing methods have consequences for future professional life (Rosser 1988). The ability to compete, which is essential in a highly competitive environment for future scientists in the aforementioned fields, and even more so for future astronauts, is negatively perceived if possessed and pursued by women (Kocum et al. 2017). The aforementioned underrepresentation of women in technical fields and professions, as well as the still negative reception of female competition, can be explained in terms of the dominance of patriarchal and sexist cultures. This culture to this day opposes the guiding ideas of second-wave feminism, that is, genuine equality and freedom of choice for women with regard to work and home, and is responsible for only partial success of the demands of second feminism (Swinth 2018).[12]

This raises serious concerns about the nature of future space missions if they are to be more massive. Our assumption here is that space technology will remain at a level similar to that of today, which will require adequate physical and technical preparation from astronauts. Thus, astronauts will probably remain a very exclusive profession for a long time, and access to it will be conditioned by fulfilling numerous physical criteria, as well as criteria concerning knowledge and skills. The latter are rather associated with "male hardiness" and male skills. Thus, if female representation in STEM remains at its current relatively sparse level, it is difficult to expect an increased presence of women in space exploration and exploitation. The gendered stereotyping of the astronaut profession represents a barrier, as do activities such as exploration and exploitation. Another barrier is the lack of easy access to space and the need for special preparation of candidates and the equipment and infrastructure.

In addition to discrimination and stereotyping, there is also the risk of sexual harassment and the perception of women through the prism of their physicality, constantly, in terms of potential sexual objects (Mason 2022, 39). This risk is particularly high in areas traditionally dominated by men and perceived stereotypically as masculine.

Exclusion of a Certain Type of Women

The problem of the exclusion of certain types of women in general, or the greater exclusion of certain types of women compared to others, is addressed primarily by multicultural feminism as well as the global and postcolonial feminism that grows out of it (Ortega 2006; Tong 2009, 200; Khader 2019; Pitts et al. 2020).[13] The problem addressed here is the different situation of women, which cannot existentially be considered one and the same in contrast to that of men. While historically women as an entire gender have been discriminated against, and this is often the case in many ways today, the opportunities, chances, and degree of actual power of individual women also vary. It is useful here to apply the standpoint and particularity perspective, understood literally, and to look at the specific situation of a particular woman. However, it is impossible to analyze individuals. What can be done is to become aware of the specific position of individual types of women, understood as groups that differ in characteristics such as, for example, origin or skin color (Nussbaum et al. 1999).

As multicultural feminism argues, for some women the fact of being a biological woman and having a typical female reproductive biology is not a major, or any, source of oppression and discrimination. Such a woman may experience oppression and discrimination because of having another characteristic (Tong 2009, 202). This, however, does not exclude the specific combination of at least two features subject to discrimination as emphasized by the concept of intersectionality, or the difficult to establish and prove, but undoubtedly occurring, specific discrimination against a woman with a given feature more than against a man with the same feature.[14] This is not the rule, however, because often the possession of a particular trait, such as a disability, can result in an equal degree of exclusion for all genders.

Global and postcolonial feminism emphasizes the different situation of women not within the same nation or community, but in the context of the whole world, especially as a result of the global division of labor between the Global North and Global South. Global and postcolonial feminism argues that while for feminists from the Global North the main concern, understood as the main source of oppression, is the issue of sexuality and reproduction, for feminists from the Global South and representing a standpoint specific to that area, political and economic issues are at the center of their concerns (Tong 2009, 215).

Exclusion of Groups Other Than Women

As I emphasized in my introduction to feminist philosophy, women have traditionally been marginalized and excluded, but they are not a monolithic group. Among traditionally excluded groups such as women, indigenous people, people with disabilities and racialized minorities, able-bodied white women are in the best position, while women whose gender overlaps with at least one other category are always at a disadvantage compared to white and able-bodied women. The problem is not only the exclusion of the aforementioned groups from direct participation in space missions, but also their exclusion from the narrative about space and our future in general. Tony Milligan speaks here of the risk of treating excluded groups, especially indigenous people, as superfluous who play no role in the creation of the new world (Milligan 2022). The discourse and thinking about human future in space should be decolonized, that is, supplemented and perhaps even replaced by thinking specific to black and indigenous people, especially on topics concerning labor and territory (Rubenstein 2022).

An example of the exclusion of racial and ethnic minorities is NASA's marginalization of African Americans. As Stephen P. Waring and Brian C. Odom illustrate in their collective work, African Americans were highly critical of the US space program in the second half of the twentieth century, which was, in their opinion, not only a waste of public money but was also designed to reinforce white supremacy (Waring and Odom 2019).

Since space exploration and exploitation requires adequate human, technical and knowledge resources, traditionally excluded groups—both within the spacefaring nation and at the level of international relations—will find it difficult to participate in space exploration. Those without adequate training and education will be excluded from enjoying the benefits that presence in space can bring. In addition to women, these groups are primarily people with disabilities and sexual and racialized minorities. It can be assumed that the quasi-militarized environment of space missions will be hostile, as will the military, to all of these groups, especially the disabled and sexually nonbinary.

Fundamentalist Capitalism in Space

An important element of the feminist critique of social relations is a critique of capitalism. Capitalism in space missions plays a key role today, since the political motivation inherent in Cold War–era rivalries is minimal, and private businessmen will play an increasingly important, probably leading role in the future. A specific threat that may arise from space exploitation is the threat known as "disaster capitalism," motivated by "fundamentalist capitalism." Fundamentalist capitalism stands for the capitalism inherent in the representatives of the Chicago school, based on profit maximization with the abolition of all barriers potentially inhibiting profit making by corporations. The US war on terror in Iraq and Afghanistan is an example of the implementation of the intentions of this capitalism. As Naomi Klein describes, the US government, which at some point lost the ability to distinguish between the good of the state and the good of the corporation, made the war in Iraq a war unto itself, aimed at profit for private companies (Klein 2007). This was a kind of privatization of the war, where the state ceded many functions traditionally performed by the state to private companies. Feminist critics often interpret this war as "robbery" and the laundering of public money in state agencies (Mann 2014, 169–170). The justification for this war required, as Bonnie Mann argues, the engagement of a powerful propaganda structure based on the concept of American exceptionalism with its essence of manhood, the defense of sovereign masculinity, and prowess (Mann 2014).

This is a situation that goes beyond the liberal philosophy of equal opportunity and equal chances. After all, not everyone has the ability to invade another country and then rebuild it as a source of income. It is therefore a unique form of exclusion that can be considered doubly immoral. The first violation of morality occurs through the very idea of taking advantage of a privileged position. Another immoral behavior is the use of a tool such as war to further one's own business. Traditionally excluded groups become even more marginalized both by the transfer of additional budgetary resources to warfare and by being direct and indirect victims of war.

Could an analogous scenario play out in space? There seems to be a strong case for pursuing the idea of fundamentalist capitalism in space exploration (Davenport 2018). It will be possible in a situation of public financing of investments in space, which will have questionable justification, and will require the involvement of private companies. Signed in 2015 by US president Barack Obama, the Commercial Space Launch Competitiveness Act gives US private companies and American citizens the right to own anything they can get in space. The act marks a repeat of the capitalist exploitation known from Earth in space. Martin Elvis calls this danger "the founder effect," which is that precedent-setting regulations created by the pioneers of space exploitation can be very difficult to change later (2022, 164). This is inevitable because of the attractiveness of space resources. Among other things, they are necessary to produce fuel from water frozen beneath the Martian surface or the moon, using carbon dioxide from the atmosphere, enabling the return from Mars to Earth. This is a possibility in the near future, basically a necessary project for missions between Earth, the moon, and Mars. A slightly more distant prospect, but one that justifies fuel production on the moon and Mars, is the exploitation of gold, silver, platinum, iridium, or rhenium from asteroids. The Moon also has helium-3, which could be exploited for use on Earth to build nuclear fusion reactors (Parsons 2020, 100–102).

The aforementioned disaster capitalism, capable of causing wars to create artificial demand, gives reason to believe that the primary motive and determining factor for space exploitation will be the capitalist drive to monopolize and maximally exploit all technologically exploitable resources in space at any given time. Like wars subordinated to profit, this model of space exploitation is available only to a small, elite group of spacefaring nations, thus excluding all other countries (Dawson 2021).

But fundamentalist capitalism in space is not just a space policy geared toward exploitation. It is also the status of workers in the space sector who will be deployed to work in space. The risk of worker exploitation may be that much higher than in the capitalist system known on Earth because of the likelihood of biomedical human enhancement technologies. These technologies, to some extent probably necessary to protect health and life in the space environment, could also be used to increase the efficiency of space workers. Although currently only the domain of science fiction literature, dehumanizing working conditions can be expected to intensify in the future due to the application of biomedical technologies, including synthetic biology (Vint

2021, 156). It is also worth keeping in mind potential reasons other than science fiction for exploiting workers in space. These reasons include remoteness from the environment, making it difficult to monitor violations of workers' rights, and the challenge of changing jobs or going on strike in an environment where the life-support and transportation infrastructure is controlled by the employer.[15]

Nationalism in Space

The aforementioned capitalism, as it often does on Earth, will also merge in space with nationalism, militarism, and colonialism. In addition to analyses by Naomi Klein and Ghosh, among others, on disaster capitalism and the dynamics of capitalism and militarism at the root of global inequality and climate change, it is worth highlighting the importance of nationalism in space exploration. Space exploration has been associated with many potential benefits and motives including, but not limited to, the advancement of science and technology and military, political, commercial, security, and prestige objectives. However, as Deganit Paikowsky argues, nationalism plays an equally important role. Each spacefaring nation, or country aspiring to that name, has its own specific goals and expectations related to space exploration, which are connected to security and economy understood in terms of national interest. Paikowsky proposes the term "technonationalism" to describe this state of affairs (2017, 76–78).

Paikowsky uncovers the essence of the space race and the space ambitions of individual countries, which are not only nationalistic in nature but also based on the dynamics of intergroup "us"/"them" rivalry. The author ends his book with a question about the uncertain future of this dynamic and whether humanity will be able to unite in space (Paikowsky 2017, 229). There seems to be a strong rationale for the pessimistic conclusion, and the assumption, that the degree of rivalry will increase in space compared to international politics on Earth.

In conclusion, no one can predict the actual dynamics of international relations in space when, in the near future, exploration and exploitation will become more advanced and the presence of space powers more visible. In a sense, humanity is not yet prepared for this at the level of international peaceful activity. Current regulations are either old (from the 1960s) or have no binding force. Space law talks about the global commons and proposes two universal principles for freedom of use of outer space and non-appropriation of space (Schrogl 2023). However, these rules themselves are at odds with each other, and those pursuing costly and risky space exploration, especially in the capitalist model mentioned above, will expect a return on investment.

A Special Case of Exclusion in Space: Exclusion in Space Settlement

The exclusion of both individual categories of marginalized people and globally marginalized nations and states outlined above concerns exclusion from the benefits of space exploitation, mainly commercial and national. They still have a chance to stay alive as long as continued existence in the next centuries with rising temperatures is possible in the most affected and hottest regions of the world. In contrast, exclusion from space settlement may result in dooming the excluded to extinction as long as space settlement is treated as a means of saving humanity from impending disaster. This section draws attention to the exclusion from the right to space settlement of a huge part of Earth's population, which, somewhat paradoxically, should be most interested in relocating to those ecological niches (on Earth or in space) where life under conditions of climate catastrophe will be easier.

The exclusion of the poorest countries most affected by global warming from space settlement concerns the consequences related to the fact that only a small group of rich countries is seriously planning or already carrying out exploitation and exploration missions in space. Exclusion here does not so much mean deliberately blocking the ability to pursue space policy, but rather the socioeconomic situation in these countries, often caused by the colonial policies of the Global North, which in practice exclude these countries from the circle of spacefaring powers. A natural extension of these missions in the future might be a settlement mission. It is natural that only those countries that are exploring space today will be able to cope with such a serious undertaking as space settlement. However, the consequence of this situation natural from the capitalist point of view, guided by the "first come, first served" rule, is to prevent the inhabitants of other countries from participating in the settlement mission.

Imagine a coming global catastrophe. Let us assume that information about the impending disaster is confirmed and humanity is informed. Only a few of the richest countries have the means to evacuate a small number of their inhabitants. It is worth emphasizing here that even saving all their own citizens will not be possible. These technological constraints may justify this otherwise unjust situation in which a huge part of humanity will lose the right to life, understood as the right to evacuate Earth. Selection criteria are unavoidable even within one's own population of spacefaring countries. However, this will not change the fact that the realization of space exploration and the preparation of space settlement will be carried out with the knowledge that in the event of a catastrophe, countries not involved in the project will be excluded.

The following analogy describes the problem well. Imagine that a small island in the ocean is threatened by the future arrival of a tsunami. Its inhabitants consist of both very rich and very poor people. At some point, everyone learns that a tsunami is coming in a few years that will kill all the inhabitants. At this point, the rich order special evacuation boats that will take many months to build. The poor do not have the necessary funds. After a few years, a few weeks before the tsunami arrives, the rich residents leave the island in boats. The poor remain on the island and die a few weeks later killed by the tsunami.

While the story above captures the problem, it does not reflect differences in scale. Earth's population will approach more than ten billion by the end of this century. Evacuating such a number is not technically feasible, and not only because of transportation options. The problem is finding the space and resources needed to support such a large number of people. Space has its limitations, correlated with the technical capabilities of humankind. Only as much space is available for us to exploit and settle as our technology allows. But space settlement need not be understood only in terms of life and death. Above all, it should be understood in terms of an ethic of quality of life. For it is conceivable that under certain circumstances, life in space may guarantee a better quality than remaining on Earth. Life in space may be more stable and safer than on a depopulated, post-catastrophic Earth. Such a supposition may seem unlikely given the extremely harsh conditions for life in space. However, under certain circumstances, such as a global catastrophe on Earth that will cause the collapse of civilization, a stable settlement with a guaranteed life-support system for a relatively small group of survivors may offer a more firmly established future.

The problem of exclusion with regard to space settlement does not only concern the aforementioned risk of actually preventing a huge part of humanity from participating in this project, arguably without even paying attention to guaranteeing adequate representation for different regions and social and ethnic groups. The discourse of space settlement itself is problematic and exclusionary. As John W. Traphagan argues, in this discourse dominated by the Western point of view, a politically and intellectually privileged small group speaks on behalf of all humanity, reproducing the Abrahamic religions' inherent depictions of humanity as a special group in need of special protection. Traphagan adds that a huge portion of humanity is unable to take part in any form of space settlement that we can imagine, due to living in poverty, as well as the associated limited capabilities and opportunities, among other things. But another reason is the cultural context. As Traphagan argues, non-Western religious and philosophical traditions, such as Buddhism, do not operate with concepts that make us think of nature in terms of colonizing and conquering it, or, in relation to humanity, in terms of striving to save the species (Traphagan 2019). Mary-Jane Rubenstein adds that we should create new narratives, using the voices of indigenous people and excluded groups to develop a new way of thinking about the cosmos and our belonging to it (Rubenstein 2022).

Thus, the concept of space settlement (in fact, space colonization) has a strongly Eurocentric character, rooted in the European belief in the uniqueness of Europe and Europeans, as well as their descendants in North America. An important component of this thinking, which originally concerned European conquest of a huge part of the world in a short period of time, is the belief in the superiority and uniqueness of the white race, which is understood to be the default and exemplary model of a human being. Progress is one of the central ideas of this worldview, progress that is Eurocentric in nature and sets development trajectories and models for the rest of the world (Mills 1997, 33–34).

Justification for Space Settlement

The rationale for space settlement is crucial for all who are interested in this project, both for the parties who can implement and participate in such a project, and for scientists, philosophers, and everyone else. But feminist philosophers and bioethicists have particularly strong reasons for posing the question of justification for space settlement. The former party interested in such a project seeks justification mainly in politics, economics, and technology. Feminism, however, looks at the context of humanity as a whole, as well as the beneficiaries, but also the potential sufferers, of this endeavor.

Taking it for granted that a space mission of the space settlement type will be very dangerous for those taking part in it due to the harsh conditions of the space environment, the very sensibility and rationality of such a project must be considered. Unlike the point of view of nonfeminist bioethics, which generally does not formulate objections to space settlement and the human enhancement potentially required for it (Szocik 2023a), feminist bioethics formulates a number of objections. Mere reference to the principle of respect for autonomy and the rule of informed consent is not enough to justify the threat to the human body and psyche, or the application of human enhancement.

Sending people on a long-term space mission, especially one that involves settling them in space, is an abuse, a kind of exploitation of, and experimentation on, these people. Such an action requires a particularly strong justification, going well beyond political justification, beyond economic and technological possibilities. Feminist bioethics could allow the concept of space settlement to be implemented only under one condition. Before preparing for space settlement, an essential requirement would be the establishment of an analogous settlement on Earth, identical in all respects, and therefore meeting the criterion of self-sufficiency, including water and oxygen production, with complete independence from any support from Earth. In an emergency, such an experiment should be discontinued, hence the need to implement it on Earth. This is a concept suggested, among others, by skeptics of human space exploration (Goldsmith and Rees 2022, 91), however not necessarily presented from feminist motives.

Exclusion in Space Settlement

Exclusion can be both caused and reinforced by the way places to live are designed. An infamous example is the Levittowns built in the United States in the 1940s. Levittowns were suburban neighborhoods of identical-looking houses that actually implemented racial segregation by barring nonwhites from living in these suburbs. Levittowns contributed to the development of the typical US ghettoization in housing patterns (Feder 2007). The idea of a future space settlement is often associated with an increased risk of being designed in a way that will cause or reinforce exclusion and segregation (Cockell 2015a, 2015b, 2016, 2022). Some structural and ergonomic features of a potential space settlement make it susceptible to reinforcing exclusion and restricting basic rights. It is worth mentioning here the necessity of staying in a closed, relatively small space, as well as the impossibility of free movement and migration. These features alone need not yet lead to racial segregation as in the case of the aforementioned Levittowns, but it is worth remembering that the Levittowns did likewise. Nor does the very structure of identical-looking houses in the suburbs necessitate racial segregation, which was brought about by the proviso prohibiting nonwhites from settling in these settlements. The naturalness and acceptability of racist views at the time undoubtedly provided practical justification and understanding for such an exclusionary approach.

We can imagine that today we would not find publicly expressed acceptance of such racist segregation with respect to a future space settlement. But the racial structure of future astronauts and space settlers may affect the racial dynamics of a future settlement in ways analogous to the racial ghettoization of habitation structures in the United States. The de facto exclusion of nonwhites in current space missions and the likely exclusion in future space settlement could lead to the racial structure of a future space base being dominated by whites—all the more so since the leader of space missions remains the United States, a country with a racist past, where racism remains a structural, institutional, interpersonal, and internalized problem. Future nonwhite astronauts and settlers in space may be excluded. Such a situation would lead to ghettoization in space. Therefore, it is important to guarantee parity and affirmative policies now, in the space exploitation currently underway. On the other hand, if we pay attention to the history of expansion on Earth, skepticism about political and social dynamics in future space settlement is justified. The idea of coloniality, as well as the fact that narratives of space exploration echo narratives of colonialism, leads us to suspect that since the colonization of Earth proceeded in a way that was far from equal and fair, the same process will similarly play out in space.

Notes

1. See also Russell 2022 on medical racism in women's healthcare.
2. Jennifer Block points out the problem of the medicalization of female biology and discrimination of women in healthcare. As Block argues, these are not the result of capitalism or medical orthodoxy, but patriarchy (2019, 8). This "medical model," applied usually to women, is characterized by an individualistic, illness-centered approach in which illness is interpreted in terms of the patient's individual problem and not as a partial effect of social construction. This represents a particularly narrow approach that does not take into account the specific social and cultural situation of women (Liburd and Rothblum 1995, 177–178). In this model, the woman, rather than social circumstances and other external factors, is held responsible for the well-being of the fetus and future child (Hubbard 1989).
3. Women's focus on weight and appearance not only incapacitates them but also makes them vulnerable to marketing manipulations centered around the diet and weight loss industry (Thone 1997). The Western obsession with thinness, especially for women, is part of an idea of beauty that is not only commercialized in capitalism (Wolf 1992) but is also part of a broader politics of racial discrimination and white supremacy expressed, among other things, by the concept of "white settler femininity" (Gentile 2020).
4. See also the so-called racial capitalism (Ghosh 2021, 118).
5. This is another paradox of patriarchal culture, where perfectly cooperative women, usually with other women, have been relegated to the private sphere, while cooperation between

men in the public sphere has often been inept, as evidenced by the many conflicts, wars, and assaults resulting from violations of social contract theory.

6. This part to the end of this section is a modified version of Szocik 2023b.

7. Feminist global ethics reminds us that it is not uncommon for women in the Global South to have a different idea of what is really problematic than women in the Global North. Therefore, as Tong proposes, feminist global bioethics should be based on care rather than rights, because the concept of rights is too abstract and does not take into account the specificity and complexity of the situation of the Global South (Tong 2022).

8. The differences between women and men are due to unequal treatment (Cameron 2007, 12).

9. The first woman to participate in a space mission as a mission specialist was Sally Ride in 1983. But she was not a pilot, meaning she was still excluded from the elite club of pilot astronauts reserved only for white men. The first female astronaut pilot was Eileen Collins in 1995.

10. As feminist global bioethics points out, the exploitation of the Global South by the Global North also leads to negative health consequences for the former (Tong 2022, 26). That is why it is so important to include the epistemology of the Global South in mainstream bioethical discussion, so as not to reinforce unjust structures and power dynamics in bioethical discourse (Ambrogi et al. 2023).

11. There are many additional reasons why women are underrepresented in advanced STEM fields and the astronaut corps, including the discouraging impact of sexual harassment in these fields and the disproportionate care requirements for women.

12. Some feminists point out that patriarchal culture in the form of global neoliberal capitalism has assimilated some of the guiding slogans of liberal feminism in order to pacify it (Ellerby 2017).

13. As Susan Sherwin argues, a multicultural approach that includes the perspective of people from other than the dominant culture is a welcome extension of the ethical and bioethical debate (2001, 183).

14. An example of the exclusion characteristic of the concept of intersectionality is black women who have been, and continue to be, excluded not only because of the color of their skin (like black men) but also because of their gender (like white women). This led to paradoxical situations in which black women did not find support from either white feminist women's movements or black men's movements, such as the Black Panthers, for whom they represented a threat to "black manhood" (Hay 2020, 11).

15. See Kendal 2021 and Nesvold 2023 on the exploitative potential of space capitalism.

4

Disability in Space

One of the most excluded groups in the types of space exploitation and exploration discussed in the previous chapters is people with disabilities. It is worthwhile to treat people with disabilities as a separate category of excluded people that demonstrates the specific dynamics of exclusion and marginalization.[1] This chapter has two main focuses. One is the assumption that space is changing our thinking about disability. In the cosmos, everyone is disabled and everyone requires special accommodation. A person who is considered fully functional on Earth becomes disabled in an environment of altered gravity. Both a person with full mobility and a person with leg paresis are equally capable (or incapacitated) in a microgravity environment.

The second assumption is the notion that the space environment is unique in terms of both risk and opportunity for including people with disabilities. The current exclusion of people with disabilities may lead to their permanent exclusion from all space activities due to the belief that the space environment is unique in its difficulty and dangers. At the same time, however, the very elite nature of current space missions creates opportunities to build from scratch new types of ergonomic structures and solutions in spatial planning and residential structures that will be devoid of the current dominance of nondisabled people. As Sheri Wells-Jensen (2023, 239) notes, disability in space in the future will be inevitable, and disabled participants in space missions will live and cooperate together with the abled perhaps in ways not known on Earth.

The ideal is to design the environment of all types of space missions in such a way that it takes into account the ability of both nondisabled and disabled people to move and operate freely. The chapter proposes a fully fitness-equal approach that precludes a situation in which the default starting point is the able-bodied person and then adaptations or accommodations for the disabled are applied.

Feminist Bioethics in Space. Konrad Szocik, Oxford University Press. © Oxford University Press 2024.
DOI: 10.1093/9780197691076.003.0005

Feminism and Disability

A major challenge to disability is the normalization paradigm.[2] In the case of disability, that paradigm is ableism. Disability can often limit opportunities in various areas of life because of ableism, or the recognition of ableism as the norm. People with disabilities may be hindered in many tasks, as well as being limited in their opportunities due to technical and infrastructural constraints. In addition to these objective difficulties, which result from the discriminatory way in which a particular environment was designed for the disabled, it is also important to bear in mind the subjective feelings of the disabled. It happens that subjective feelings are caused, or at least reinforced, by objective circumstances that cause or reinforce the exclusion of the disabled. A particular case is the decision to die, to euthanize, or to discontinue life-sustaining treatment taken by people with disabilities. Ableism can cause such decisions to be driven by a belief in the low value of one's own life. This case also shows how important it is to create the right healthcare conditions to enable vulnerable people to make autonomous decisions that are based on their beliefs and values, and not determined by a lack of, or limited capacity for, care (Ho 2014, 346–349).

Disability bioethics, like feminist bioethics, analyzes the specificity of being in the world as an alternative embodiment. In a sense, both bioethics are focused less on female embodiment or impaired embodiment than on the definition of normality (Scully 2008, 176–177). In the case of feminist bioethics, male embodiment is considered the norm with which feminist bioethics argues. In the case of disability bioethics, such a norm is the able-bodied body. Disability, like gender, has both a bodily context and a dimension of social construction (Scully 2022, 182).

Disabled people's situation may deteriorate depending on the social identity they have, regardless of their disability. Certain identities perceived negatively in many cultures, such as being female, then being a woman of color, lesbian, or transgender, overlap with negative social perceptions of disability and compound the discomfort of the person with a disability. In the case of women with disabilities, we can talk about attitudinal barriers that are generated by the cultural identification of women with socially deprecated characteristics, such as childishness, emotionality, and weakness, which compound the disadvantages of disability (Saxton 2016, 42).

Feminist bioethics rejects the medical model of understanding disability, according to which disability is the domain of the individual, not society, and the solution to disability lies in individual treatment, not modification of social structures (Jenkins 2016). However, this does not change the fact that some disabilities under certain circumstances can be something that is disadvantageous to a disabled person. This is in line with the value-neutral model of disability proposed by Elizabeth Barnes. Barnes interprets the disadvantages that can occur due to disability as disadvantages caused by the fact of being in a minority, but she supplements this mere-difference model of disability with the aforementioned value-neutral model, in which for some people in some circumstances—or any circumstances—being disabled would be something bad even in an ideal society (Barnes 2016, 79–80, 89–90). Extreme environments such as military and space missions exploit this component of the inconvenience of having an impairment.

Feminist disability scholars point out that despite the disadvantages in an environment dominated by and designed for nondisabled people, disability can be a form of empowerment and constitute a positive, minority identity (Burch and Kafer 2010). Some feminists, such as Mia Mingus, criticize the ableist paradigm, which has a total impact on all other identities, not just those of people with disabilities, but all other bodies that do not meet the criterion of functionality set by the norm of a white, male and able-bodied body (thus the bodies of women, immigrants, or trans people) (Mingus 2017). Important to feminism, the concept of interdependence is also used by disability scholars to show that, in essence, every person needs care, is dependent on others, and expresses a different form of existence and corporeality (Withers and Ben-Moshe 2019). Disability researchers also point out the usually unconscious link between able-bodiedness and heterosexuality, where both able-bodiedness and heterosexuality in practice function as compulsory. Indeed, in a heterosexual, ableist culture, there are not only norms of fitness and bodily beauty, but also of pleasure, including sexual pleasure, that exclude, or at least discriminate against, both disabled and queer people (McRuer 2006).

Able-bodiedness and Performance in Space Missions

From what we have said so far, what emerges is a vision of space, or rather the politics and philosophy of space exploration as a place that is hostile to all who are not able-bodied (white) men. Space is seen not only as masculine (not for women), but also as requiring exceptional health and fitness (not for the disabled). Excluding the disabled from participating in space missions may have a practical, environmental rationale, but may also stem from stereotypes.

The feminist critiques of standard understandings of health, illness, and disability as largely socially constructed categories are often justified. The history of medicalization and pathologization of conditions that we today recognize as natural human variation is a prime example. We agree with Kristina Gupta, who proposes that we take a neutral, subjective, and contextual approach to the categories of illness and disability that takes as its criterion their subjective impact on the quality of life of the individual concerned. In this approach, the same or a similar condition for one individual may be neutral, for another beneficial, and for yet another vexatious and in need of intervention (Gupta 2020, 118).

However, what is acceptable and understandable with respect to so-called normal situations and environments becomes challenging with respect to extraordinary or exceptional situations. While we can acknowledge that our environments are typically constructed with the concept of the healthy (male) body as the norm and model, and thus unfairly exclude those traditionally labeled sick and disabled (and women), it is unclear how to treat exceptional environments. Such environments include, for example, the military and the space mission environment. As suggested elsewhere (Szocik 2023), the environment on Earth most twinned to the space environment in terms of ethical specificity is the military mission environment and military ethics. From a feminist perspective, it is unfair that access to the military is often restricted to those deemed fit and healthy. It is conceivable that some of the tasks performed today by the physically fit soldier could be performed by a person with a disability, and any differences in performance between the two would be bridged by technology. However, there are possible battlefield scenarios in which not all tasks will be able to be performed by technology, which may not always be able to replace humans or compensate for the difference in performance and ability between the able-bodied and disabled soldier. It is also possible to have a scenario where the technology used by the disabled soldier no longer works. Such technology may become a deliberate target of an enemy army, which may obtain knowledge of the use of such technology and deliberately seek to neutralize it, thereby weakening its opponent's army. This does not exclude the possibility of electronic neutralization of any technology, regardless of the soldier operating it, which weakens the strength of the aforementioned argument. Thus, this would be at least one possible scenario in which, as a result of the loss of this compensating technology on the battlefield, a fit soldier would have an advantage over an unfit one. This example illustrates the nature of exceptional environments, of which the battlefield environment is undoubtedly one. In a sense, the space mission environment is also such an environment. The compulsory able-bodiedness that exists in an oppressive, ableist society (McRuer 2006) is especially strongly replicated in extreme environments such as military and space environments. This culture of compulsory able-bodiedness is particularly acute in such environments, making it difficult to objectively separate authentically demanding conditions from the social ideal of bodily perfection.

Let us assume that there are types of space mission and potential means of space travel such that people with disabilities will not be excluded. The ideal situation from the point of view of equal access would be one in which any individual could participate in a given type of space mission. At the same time, a solution that would not interfere with this ideal of equality would be one that allowed for optional medical intervention if the mission organizers were unable to develop the technology to allow disabled individuals to participate in a given type of activity. This is a hypothetical situation that assumes that no moral or legal responsibility can be assigned to the organizers for intentionally or unintentionally excluding people with disabilities. It is a situation in which there are objective factors that make it possible for only an able-bodied person to accomplish a given task in space, at least with respect to a given characteristic. The battlefield analogy for this situation is the objective requirement that a sniper have eyesight in a situation where there is no technology that can enable a blind person to act as a sniper. As long as the technology to replace human vision has not been developed, we can assume that excluding blind people from recruitment for the position of sniper does not violate the ideal of equality.

Another problem is the normalization paradigm operating in the military community and defense policy, which may determine the preference for a particular type of military technology research. This normalization paradigm, if indeed it exists, will preclude research designed to enable the involvement of people with disabilities on the battlefield. Thus, regardless of whether a particular type of technology to compensate for a certain kind of disability on the battlefield is technically possible today or in the future, we may never know because research on that technology, as incompatible with the normalization paradigm in the military, is not even undertaken. Consequently, this fact that indeed under current circumstances a given disability excludes the person who has it from performing a given function on the battlefield (i.e., a blind person cannot become a sniper active on the battlefield today) does not change the fact that there is an exclusionary context that does not even allow for the possibility of engaging people with disabilities in certain activities on the battlefield and does not even take any action (among other scientific and research work) to invent technologies to compensate for the disability.

With respect to space missions, the equality ideal requires that mission organizers recruit people with disabilities today, prepare candidates with disabilities to perform particular tasks and duties, and, if necessary, equip them with the appropriate technology to perform those tasks that are believed to be performed only by nondisabled people. Furthermore, mission organizers should work to modify the infrastructure and make it accessible to persons with disabilities. The risk with regard to applying the ideal of equal access to the space mission environment is that its difficulty and uniqueness may be exploited and considered objective and insurmountable. Opponents of open access to participation in space missions may justify their position by arguing that the space environment and space flight are themselves difficult and demanding, requiring not only astronauts with an abled body, but astronauts with exceptionally high fitness and perfect health. If the concept of human enhancement were to be applied to space missions, its existence may strengthen this argument by suggesting that even "abled" and healthy individuals are excluded from participating in space missions unless they are specifically enhanced through biomodification. The problem that characterizes the space mission environment, therefore, lies in its specifically demanding nature, which means that it can be difficult to distinguish the actual, objective difficulties that potentially exclude people with disabilities from the ideal of ableism, which means that we may tend to find it natural and acceptable to exclude people with disabilities from participating in space missions (and other demanding environments on Earth).

As Stephanie C. Jenkins (2016) argues, ableism dominates both bioethical discourse and theories of morality and personhood, in which moral status and personhood are defined according to performance criteria. A disabled person who does not meet these performance criteria is therefore an incomplete type of person. The space mission environment is associated with a requirement for above-standard performance, which is served, among other things, by the concept of radical human bioenhancement considered in space bioethics. In this light, the environment of space missions appears uniquely ableist and uniquely closed to people with disabilities because of the central role played by the category of performance for bioethical and moral theories.

A paradigm shift in the way we think about space exploration, as well as in space policy, should incorporate the concept of universal design, the essence of which is to design a given environment as inclusively as possible so that it allows the widest possible range of users to adapt to it (Wells-Jensen et al. 2019). As Wells-Jensen and coauthors argue, a universal design would not only benefit people with disabilities but also able-bodied crew members in the event of an emergency when certain systems stop working. The authors also discuss advantages potentially possessed by blind astronauts over sighted ones, such as no risk of vision loss from cosmic radiation exposure or less difficulty orienting in microgravity. Another important element raised by the authors is the likelihood of some able-bodied mission members acquiring disabilities. Due to the nature of space missions, which require the fitness and commitment of all mission members, a person with an acquired disability will be required to return to mission tasks (Wells-Jensen et al. 2019). As Christiane Heinicke and colleagues point out, mission planning should be created with, not for, people with disabilities (Heinicke et al. 2021). But engaging people with disabilities requires knowledge of their behavior and reactions in space, and this knowledge, given the exclusion of this group from space exploration—as well as the marginal participation of other traditionally excluded groups such as women and people of color—and the focus on non-disabled professional astronauts, is minimal (Marge 2022).

Can It Be Justified to Exclude People with Disabilities from Participating in Future Space Exploration?

The European Space Agency's classification includes three categories of access to participation in space missions for people with disabilities: the red category (prohibition of participation due to inability to perform the task), green (admission), and yellow (admission after applying modifications to the space mission) (Gres et al. 2022). The ideal situation from the perspective of feminist bioethics would be to apply the green category to future space missions.

The main threat to the inclusion of disabled astronauts in long-term space missions is the concept of biomedical human enhancement. The idea of human enhancement is a challenge to disability in general, including on Earth. But in the case of space, because of the demanding environmental conditions—which, however, may not actually be particularly challenging for people with disabilities, but rather be taken as an argument for their exclusion—the participation of people with disabilities would interfere with the idea of human enhancement, based on human improvement. But even if it were decided under such circumstances to accept the participation of people with disabilities in the mission, there would still be a risk of seeking to offset their disabilities through human enhancement measures. Consequently, a space exploration scenario based on mandatory, or at least widespread, application of radical human enhancement would be an exploration scenario that excludes people with disabilities as outliers in the profile of the ideal participant in long-term space missions. The logic underlying the concept of human enhancement seems to be at odds with disability in the context of space exploration, at least as long as space exploration and exploitation is possible for a small group of highly capable and trained participants. The massification of long-term space missions may lead to increased inclusivity and to a lessening of the link between the logic of human enhancement and disability. In the case of larger-scale space exploration, human enhancement could be utilized more for therapeutic purposes for people deemed disabled for space exploration. It would also be important to define what constitutes a disability in terms of the specifics of space missions.

In the approach to feminist bioethics proposed in this book and the critique of ableism in relation to the space mission environment, as we formulated it above, a significant threat in the near future is the idea of human enhancement. This is undoubtedly an underrecognized type of threat today, but if the development of space medicine proceeds in the direction of applying increasingly sophisticated and invasive biomedical modifications to future astronauts, people with any disabilities may be further excluded as a result of the culture and politics of compulsory biomodification. Here again, the broader context, not only strictly bioethical, but also social and political, which feminism points to, is useful. The space mission environment is masculinized and associated with a macho cult. Impairment is something that opposes such a notion of space as a place meant for above-average, able-bodied people, usually (white) men. The presence of people with impairment in this environment is not accepted. However, the problem with excluding people with impairments from participating in space exploration is different from their exclusion in other environments on Earth, where the situation is improving and remedial measures are being carried out. In the case of excluding people with impairments from space missions, their exclusion is not merely preventing them from performing a particular type of occupation, which might be more understandable under certain conditions, as in the case of people without impairments who nevertheless do not meet certain conditions for performing particular types of work. However, participation in a space mission cannot be reduced to only performing the profession of a pilot, mechanic, technician, or mission specialist. We are talking about enabling humanity to live in a new place from which the disabled are excluded. Consequently, excluding an impaired person from the possibility of becoming an astronaut is not just depriving her of one of her professional job opportunities but is tantamount to excluding her from the possibility of living in a new environment.

Long-term exploitation and exploration of deep space potentially leading to permanent settlement in space therefore carries the risk of permanent exclusion of people with disabilities, but also of any others diagnosed during the selection process as unsuitable for the restrictive requirements. Kelly C. Smith and Caleb Hylkema raise the problematic nature of the demand for full inclusivity of space exploration, which would be open to all types of disabilities. They point out that the high cost of such demanding missions, at least in the initial phase dedicated to a small number of participants, will require the full involvement of everyone, which will exclude the participation of people with impairments who themselves require the care of others (Smith and Hylkema 2020, 228). Their point of view sounds rational, but it is worth remembering that the constant context for planning space exploration is the possibility of building a permanent human settlement in the future, which will perhaps be a kind of refuge for at least part of humanity. The rationale for this space refuge idea grows over time with the advancement of space technology, as well as the deterioration of living conditions on Earth.

This pragmatic view of space exploration combined with the idea of human enhancement will probably reinforce ableism in the space policy with regard to candidate selection. If the goal of human enhancement were to make astronauts even better in terms of their health, resilience, strength, fitness, and performance, any person with a disability would be treated as being below the norm inherent in a space environment with a macho culture. With regard to people with disabilities, the idea of human enhancement, applied to participation in space missions for fairly common-sense reasons, could lead to the future exclusion of people with disabilities and the creation of a very exclusive society in space. The pragmatic argument based on the high cost of modifications for people with disabilities is undermined by the fact that, as Wells-Jensen notes, in space we can start building everything from scratch, taking into account the needs of all people, which will significantly reduce costs, while increasing comfort for all (Wells-Jensen 2022).

In conclusion, the feminist bioethics of space exploration proposed in this book opposes the exclusion of people with disabilities. In addition to arguments from the critique of the normalization paradigm and the ambiguous consequences of human enhancement, as well as social justice, an important argument is the expected future nature of space exploration, which may culminate in the concept of space settlement and space refuge. Excluding people with disabilities would therefore not only mean denying them the right to certain types of activities and professional work, but under certain conditions could be tantamount to denying them the right to life and survival.

Negative and Positive Selection for People with Disabilities in Future Space Settlement

The possibility of allowing people with disabilities to participate in space missions, or the policy of excluding them, has important implications for the shape of future space settlement. This is because it is a question of the vision of the future new human population in a permanent space base, which will—or will not—include people with disabilities. The exclusion of people with disabilities in the initial stages of space missions, taking into account those currently being carried out in Earth orbit, may in the future justify their exclusion from more advanced and common types of exploration missions, including the concept of space settlement, treated under certain conditions in space refuge terms. The selection of disabled astronaut John McFall—however, a Paralympic medalist, a white British male—by the European Space Agency is a welcome development, while it is difficult to say whether it will entail fundamental structural changes in the way we think about accessibility to space missions (John McFall, n.d.).

But the openness of space policy to people with disabilities is not just the selection of future crew, discussed in the sections above. It is also a eugenics policy for reproduction in a future permanent or semipermanent space base. It seems that quite popular intuitions about disability favor negative selection—or eugenics—which means preventing bringing people with disabilities into the world. One such intuition is that disability-free intuition growing out of the principle of procreative beneficence may lead one to think that we should avoid bringing people with disabilities into the world. Another intuition is expressed by the so-called standard view, according to which having a disability is always, or almost always, a bad thing for the person with it (Mosquera 2022, 590–591, 594). Assuming the scenario of a future space base in which humans will reproduce, what status should be accorded to the concepts of negative and positive selection with respect to traits diagnosed as disabilities?

Feminist bioethics tends to take a negative view of the concept of negative selection, which is synonymous with genetic determinism. This determinism reduces the quality of life of a future human being based on established—or not—genetic traits detected in an embryo or fetus (Hall 2013, 128). Feminism rejects this model because, for feminist bioethics, what is important is the social and structural context in which a woman makes the decision to carry a pregnancy or abort it, as well as subsequently to have a child with a disability. The structural and systemic consequence of genetic determinism based on the principle of reproductive beneficence can be the marginalization of the status and needs of people with disabilities. Biomedical intervention through gene editing or negative selection resulting in the abortion of fetuses with detected defects may be the only acceptable solution in such a model.

The development and popularization of genetic research may have negative consequences for space exploration in the context of inclusivity. As Nesvold (2023, 47) notes, genetic screening of candidates for space missions may focus on arbitrarily selected traits deemed desirable for the future population to inhabit space, and consequently lead to eugenic policies. The slippery slope argument seems to apply here. Once the idea of genetic screening and selection for potential genetic defects is introduced, the practice could become a mandatory policy in space missions.

Feminist bioethics also rejects the concept of positive selection. Such a policy for the space mission environment will inevitably lead to the total exclusion of people with disabilities, because by creating such a policy for space exploration, humanity will replicate the patterns on Earth, which is, after all, an easier place to live than the demanding space environment. It is therefore important to pay attention to the goals of the space mission and its beneficiaries. As long as such a mission is to be political, militaristic, and commercial in nature, the potential acceptance of both negative and positive selection as models of reproductive politics can be expected. Both models will exclude people with disabilities as being of no use to the militaristic and capitalist model.

In conclusion, the concepts of negative and positive selection, well known on Earth and often supported by mainstream bioethicists who cite the well-being of the future child and sometimes society, may pose a serious challenge to the way space policy is organized—all the more so as we can expect even more medical developments, especially for space exploration. Feminism applied to these considerations of future space missions leads to the conclusion that seemingly common-sense ideas—the well-being of future participants in space missions, as well as perhaps children in the future born in space settlements—should not be pursued through policies that eliminate all people with any disabilities. Such policies should be focused on structural change, not individual responsibility. If, in line with Smith and Hylkema's idea mentioned above, at least initially, all participants in space missions will be needed to work, we should change the way we think about space missions, about work, and about caring for others, and include in such missions from the beginning people dedicated to caring for others. The environment of space missions is not only an environment of strong and reliable people but also one of weak and disabled people. And caring for others is not a vocation or a sacrifice, but the same job as that of a pilot astronaut, mission specialist, or engineer.

Notes

1. The common core of feminism and disability studies is a critique of the idea of the norm, as well as a concern for the marginalized (Mosko 2018, 1369).
2. Normalization would not be possible without linguistic categorization, which serves to reinforce the binary identities that dominate a masculinized culture (Angouri 2021).

5

Feminist Bioethics of Human Enhancement

A Feminist Critique of the Normalizing Paradigm in Medicine

Human enhancement refers to the improvement/modification of the human body and/or psyche through biomedical interventions to a level beyond the requirements of staying alive and healthy (Juengst and Moseley 2019).[1] In a sense, the concept of human enhancement deserves to be criticized doubly by feminism. The first object of feminist criticism is the concept of normalization. Feminist bioethics criticizes the paradigm of normalization in medicine, which presupposes a match with the "norm." Criticizing this paradigm can lead to at least an apparent conflict when what is criticized is at the same time desired by the parties concerned with treatment themselves (Gupta 2020, 2). The concept of health and illness is based on a certain understanding of the norm. Depending on how one defines health or illness in a given circumstance, particular groups of people can be excluded.

The aforementioned conflict between the individual's desire to obtain a medical intervention associated with improved individual well-being and the critique of the normalizing paradigm in medicine leads to a kind of collective action problem into which excluded groups unwittingly fall. Any medical intervention that enhances the well-being of an individual representing a marginalized group increases that person's chances of autonomy and authority in an oppressive society, but at the same time deepens or reinforces the existing system of oppression in which the individual must conform to an identity imposed by the dominant narrative.

The second object of feminist criticism is that the concept of human enhancement establishes a new understanding of the norm. Although the traditional understanding of human enhancement implies transgression of the norm, for proponents of human improvement it is this transgression that becomes the goal, and thus the new, exclusive norm. The concept of human enhancement doubly invokes the notion of norm and the paradigm of normalization, and as such becomes the object of feminist critique twice over.

Feminist Bioethics in Space. Konrad Szocik, Oxford University Press. © Oxford University Press 2024.
DOI: 10.1093/9780197691076.003.0006

However, while the traditionally marginalized and excluded have a chance of being included in this first understanding of the norm as it operates in medicine and healthcare, they almost certainly have no chance of being able to match this specifically understood norm in the human enhancement paradigm.

A feminist critique of human enhancement does not necessarily explicitly presuppose the concept of perfectionism. As indicated elsewhere (Szocik 2023a), I do not believe that the concept of perfectionism is at the core of the concept of human enhancement. Nevertheless, the former becomes useful for feminist critique. Human enhancement functions as an alternative vision of the world and society in which people possess, or should possess, according to a new conception of the norm, new or at least enhanced qualities in comparison to the rest of the population. This "rest" is usually people who are already excluded and marginalized. These people are perceived as inferior, less resourceful, less intelligent, less educated, and usually also morally inferior. Human enhancement, which is supposed to improve all these functions, can further stigmatize these groups. Thus, the concept of human enhancement contains a certain idealization of normalization, a new normalization that is at least implicitly defined in contrast to those characteristics that are currently considered normal or average. But even if we were to decide to assume that the ideal of perfectionism is inherent in the idea of human enhancement, the environment of space missions, due to the harsh conditions, leads to a situation in which every modified participant in such missions—assuming they all undergo biomedical modifications for environmental reasons—is indistinguishable from the others. The only difference, therefore, between those who have reached the ideal of perfectionism and those who have not been modified would be between the population living on Earth and those on long-term space missions.

Some feminists criticize the distinction between health and illness, which they see as a manifestation of social and political oppression (Gupta 2020, 5).[2] Moreover, the idea of human enhancement's assuming a division between therapy and enhancement is an even more extreme example of an oppressive approach. The concept of "livability" proposed as an alternative to the health/disease distinction may give the concept of human enhancement some chance of acceptance under certain conditions. It seems, however, that from the feminist point of view, it is much more difficult to justify the moral legitimacy of human enhancement than that of medical interventions referring to the categories of health and illness, even if in both cases the currently used categories are replaced by the term "livability." Feminism questions or simply fails to recognize the distinction between therapy and enhancement that is fundamental to the concept of human enhancement. This distinction requires a prior definition of the concept of norm, which in feminist terms is a social construction.

Another reason why human enhancement comes under criticism more strongly than the very practice of medical therapies and interventions in current healthcare is that inequality and structural oppression and discrimination exist. Medical care favors men over women, as well as white people over people of color (Gupta 2020, 25–26). The concept of health is a tool of power; it is racialized and gendered (Grigg and Kirkland 2015). This is evident in guaranteeing better access to advanced and expensive medical procedures for white men than for white women and people of color. Imagine if human enhancement, whether on Earth or for space missions, were to become publicly funded and applied equitably to all candidates for a given type of mission. It can be assumed that any human enhancement program would be quite expensive. If we pay attention to the discrimination and inequality regarding gender and race in the standard allocation of medical resources, we can assume that discriminatory structures and biases will be all the more active in the case of human enhancement, which is something optional and more prestigious and opens up new possibilities—on Earth and especially in space. Since standard medical practice is a highly discriminatory system, emergency or state-of-emergency medicine, which any medicine offering human enhancement would be, will be all the more discriminatory because of the higher stakes.

As Gupta points out, medicine is the cutting edge of normalization. Normalization in medicine means that certain states that deviate from the norm are pathologized as disease states and, as such, require treatment. Such procedures also have financial incentives. Normalization follows the biases of gender, race, or disability (Gupta 2020, 28). The negative effects of normalization, particularly acute for those already most excluded and marginalized, are magnified by medicalization in that all problems experienced by the individual in which medicine is interested are explained in terms of problems of the body and mind. This approach ignores the social context (Gupta 2020, 29).

An analogous mechanism operates in relation to the concept of human enhancement. Human enhancement, especially pharmacological and above all genetic enhancement, must presuppose some kind of biological and genetic determinism, since it offers medicalization and biomedical procedures as remedies and means. This objection can be applied to all considered applications of human enhancement, but it seems that the concept of intelligence enhancement is particularly controversial. Leaving aside the questionable medical feasibility of such enhancement, the very concept of improving someone's intelligence is exclusionary and discriminatory. It must also operationally presuppose some standard of intelligence that a person with financial resources and access to technology can hypothetically improve, ignoring the "lesser" intelligence of the rest of society.

With regard to both conventional medical interventions and human enhancement, the paradigm of normalization, medicalization, and individualization attempts to shift the burden of responsibility to the individual, without paying attention to possible, at least partial, environmental conditions inherent in, for example, structural discrimination against certain groups. Significantly for feminist critics, the ideals considered exemplary and hypothetically achievable through enhancement may reinforce racist and sexist social structures (Tong 1997, 240–241). The concept of human enhancement is more socially constructed than definitions of health and disease. As such, it is more likely to be a realization of ideals of perfectionism that are often oppressive and discriminatory.

The problem with the concept of human enhancement arising from the need to invoke the controversial notion of norm is twofold from the perspective of feminist bioethics of space exploration. One challenge is the specificity of the space mission environment, which forces us to redefine the terms "enhancement" and "therapy." In the space environment, everyone is below what can be considered the norm on Earth.

A second conceptual challenge is the social constructionism and contextual dependence of the terms "norm," "health," and "illness," emphasized especially by feminist bioethics. The criterion of the norm depends, among other things, on the elements being compared, and power structures determine how dominant a particular group becomes. Finally, the impact of gender bias is enormous, exemplified by the medicalization of those processes that are not typical of the male body that constitutes the medical model, namely menstruation, pregnancy, and menopause (Lorber 1997, 2). As Judith Lorber points out, thinking in terms of a biomedical model of the human body understood as universal should be replaced by social bodies understood as diverse and varied (Lorber 1997, 102). Women have often been excluded from many medical tests both because of their menstrual cycle, which was seen as interfering with test results, and the possibility of pregnancy in terms of potential risks to the fetus. This focus on the well-being of the fetus rather than the pregnant woman also reveals a sexist view of pregnant women as fetal containers rather than autonomous persons deserving respect because they are persons. Consequently, the medical model was also the male body in relation to how women used drugs (Rothman and Caschetta 1995, 66). The exclusion of women from examinations and tests, reinforced by stereotypical ideas about women and the tendency to minimize and ignore their disease symptoms, but also by genuine ignorance about their actual health status, leads to the exclusion of women from the benefits offered by modern technology and medical knowledge. Despite the progress in medicine, women often experience difficulties in obtaining an appropriate diagnosis and subsequently appropriate medical care (Dwass 2019).[3]

It is also worth bearing in mind the fundamental difference between feminist and nonfeminist approaches to the issue of genetic enhancement. The perspective of feminist bioethics based on categories of exclusion and posing the question of who gains and who loses in the case of human enhancement is absent from nonfeminist bioethical analyses. It is worth citing here the analysis of Jon Rueda, which is also useful from the point of view of the bioethics of space exploration, which discusses the concept of radical genetic enhancement in terms of the potential annihilation of the *Homo sapiens* species for the purpose of enabling the directed evolution of posthumans. This is an otherwise rational idea, for such is one of the main intentions of the idea of radical human enhancement, to create a better human being. Rueda devotes most of his attention to analyzing the abstract duty of beneficence, referring to the utilitarian postulate of providing well-being for as many people as possible, in this case future posthumans (Rueda 2022). What is striking about his paper is the absence of any reference to traditionally excluded groups, especially residents of the Global South—after all, the concept he discusses of radical human enhancement and the possible evolution of a new species of posthumans is supposed to apply to all humans. The perspective of feminist bioethics adopted in this book does not consider abstract principles such as the principle of beneficence or various moral duties but analyzes the ethical status of human enhancement in terms of exclusion and the dynamics of power structures.

The Social and Ideological Construction of Science

In addition to the problematic and discriminatory concept of norm, discussed above, the concept of human enhancement considered from a feminist perspective also generates methodological problems. Science is constructed and shaped by society and by ideologies, but this does not apply only to the classical understanding of the evolution of scientific discoveries proposed by Thomas Kuhn (1962). Feminism points out that the understanding of women, prior to the development of particular theories in science, plays a special role in the development of science. One example of this overlap between social perceptions of women and the direction of the development of scientific concepts is indicated by Diane E. Eyer's history of the concept of bonding between mother and infant. Despite its lack of scientific basis and critique, the concept was very popular due to its reinforcement of already widely held societal perceptions about motherhood and the special role of the mother. One consequence of this methodological blunder was that it ignored the real causes of family difficulties and child development problems. Their sources were not sought in institutions, society, or economic problems, but solely in women's behavior (Eyer 1992, 98–99). This example demonstrates at least two negative, highly socially impactful processes. One is the influence of biases and stereotypes, which shape methodologies and ways of doing science. The second is the aforementioned negative social consequences, which may consist in ignoring real causes by misdiagnoses caused by prejudice and discrimination.

The concept of human enhancement may be analogous. Depending on the target group, as well as the type of enhancement considered and the accompanying justification, individual target groups can be specifically understood through the prism of stereotypes and perceptions about them, but also their previous social position. Human enhancement policies toward women can be shaped around perceptions of women as mothers, wives, housewives, but also as a gender traditionally perceived as physically weaker, with stereotypically distinct, supposedly female-specific psychological characteristics. Similarly, such a situation may concern all groups stigmatized around such categories as race, class, disability, and others.

In the case of women, the medicalization of their lives has typically centered around their reproductive biology, as well as the belief derived from that biology in a traditional, sexist, and patriarchal society that their obligations to children and family exist. Both childbirth and the specific manner in which the newborn developed, as well as the mother's relationship to the newborn, became medicalized and were judged in terms of the norm or deviation from the norm by doctors. As Eyer suggests, medicine first defined the nature of patients according to market demands and then offered solutions to problems created by medicine itself (1992, 129–130).

This particularism and bias that characterizes science raises questions about the possibility of objectivity and universality. Feminist philosophers usually fundamentally question the possibility of an objective and independent science. One of the exceptions in feminism is liberal feminism, which allows for the possibility of science in the positivist sense. Liberal feminism assumes that biased scientific findings are the result of methodological negligence on the part of the scientist, not due to the very essence of the scientific approach, which with proper care is likely to present objective, impartial, and value-neutral results (Rosser 1994, 129–130). The problem, however, is social and power structures. Existentialist feminism points to the identification of female biological sex with inferiority in sexist social structures, although this is not based on biology. Consequently, as existentialist feminism suggests, science done in a sexist society will be sexist as long as sexist social structures are maintained (Rosser 1994, 136).

It is hard not to agree with Inmaculada de Melo-Martín's observation that from the perspective of nonfeminist bioethics, science, especially with regard to genetic modification technologies, is usually assumed by default to be axiologically and gender-neutral (de Melo-Martín 2022a, 281). The thesis of the value-neutral nature of science means that science by itself does not have the potential for negativity and harm, but this is only due to its misapplications, human intentions, or faulty policies.

Like science, disability is treated through the prism of social constructionism, but also its negative, oppressive effects. Feminist attitudes toward disability have much in common with feminist attitudes toward medicine, even when it concerns able-bodied people. In both cases, one of the central categories criticized is that of disease, pathology, and disability, whose common feature is that they stigmatize the person to whom the disease or disability is attributed. A person classified in one of these categories by the current social consensus has an inferior status, deprived of something basic yet essential to life—for example, being able to work, earn money, and support oneself. The negative status of both illness and disability is reinforced by the cult of the body and health, at least in Western culture (Gupta 2020, 115–116). In this context, the concept of human enhancement appears as an expression of a particularly powerful normalization—powerful in the sense that it goes beyond the normalization typical of conventional medicine. Human enhancement can be seen as an expression not only of disability—this nonacceptance is expressed by medical interventions aimed at normalizing disability—but as one of nonacceptance of "normality" and physical mediocrity. Feminist disability studies reject the identification of disability with deficit, criticize the total category of the normate, as well as demonstrate the interplay and overlap of oppression caused by gender and disability (Hall 2011). Shelley Tremain offers an interesting development of the feminist paradigm of understanding disability as an apparatus of power relations, which is presented as something naturalized (Tremain 2019).

Donna Dickenson argues that modern medicine, through pharmaceutical and genetic research, human enhancement, and all those situations in which some parts of the human body are collected for the purposes of medical research, treats people as objects rather than subjects. This process was known long before biomedicine and was the domain of the experience of women, whose bodies were objectified by men and for men. In modern times, as Dickenson argues, men, on a par with women, are the subjects of biomedical objectification, which Dickenson illustrates using the category of the feminized body (2017, 1–2, 8–9).

An additional problem may be the large-scale use of digital health technologies. It is quite likely that their use will be more widespread and advanced in space. While digital health offers advantages in terms of access and speed, including those attractive from a feminist perspective, such as empowerment of women and others traditionally excluded, it also carries some risks. One of these, as Alexis Paton argues, is paradoxically not empowerment but the risk of disempowerment of women, at least with regard to digital health concerning pregnancy. As Paton suggests, digital health runs the risk of shifting responsibility to women, and in addition may reinforce patriarchal stereotypes and expectations about women's role in reproduction (Paton 2022).

Human Enhancement and the Autonomy of the Female Body

As we have shown, the idea of normalization associated—however unessentially—with the concept of human enhancement is at odds with the ideas of feminism. Nevertheless, the bioliberal approach to human enhancement may be an ally for women in their efforts to regain control over their own bodies, which in a patriarchal culture are usually viewed through the prism of their reproductive functions. This association of women with reproduction has led to a situation in which pregnancy has been given the status of an abstract process, and the fetus is presented, especially by opponents of abortion, as an abstract entity developing in a space unlimited by the woman's body (Shildrick 1997, 25).[4] The liberal approach to human enhancement assumes the autonomy of the individual interested in biomodification. By analogy, the arguments supporting the right to human enhancement can be extended to the situation of a woman enslaved by the patriarchal notion of reproduction as a process independent of the woman.

Technology in a nonideal world can be a tool of oppression. We sympathize with those feminist ideas rooted in the philosophy of Shulamith Firestone and xenofeminism, which are essentially characterized by optimism and hope in their approach to the development of science and technology. Nevertheless, it is worth bearing in mind the social conditioning of technology and the interplay of society with technology and vice versa, as Hester expressed: "Technology is as social as society is technical" (2018, 11).

Like the ambiguous case of digital health technologies, ARTs can not only serve to strengthen women's autonomy, but at the same time increase the extent of third-party control over a woman and her reproductive processes. Moreover, ARTs can generate pressure to reproduce or make additional reproductive efforts by providing technologies where previously a woman (or couple) had abandoned reproductive efforts as a result of being deemed infertile (Turkmendag 2022). The dominant position in feminist bioethics is a negative assessment of the liberating potential of reproductive technologies for women. Their main drawbacks are that they reduce women to fetal container functions and make their pregnancies a matter of public record. This is due, among other things, to the objectification of the fetus, which becomes a kind of independent agent that is—instead of the woman—the main object of interest and protection, and at the same time of permanent surveillance and intervention, on the part of healthcare and society (Samerksi 2015).

The belief that a woman's lifestyle during pregnancy has a significant impact on her future offspring can lead to the reinforcement of the image of women as constantly responsible for the well-being of future generations. Such a woman, in an extreme version of epigenetics, is under constant scrutiny and monitoring by modern science. As Ilke Turkmendag notes, this paradigm fosters the instrumental use of the idea of a "good mother" as a tool to control, manipulate, and discipline women (Turkmendag 2022). The level of inspection, scrutiny, and expectation to be sacrificed by women is increasing due to the development of reproductive technologies (de Melo-Martín 2021, 292–295).

In conclusion, both human enhancement and ARTs have the potential to increase women's bodily autonomy and are certainly worthy of positive evaluation. Nevertheless, in a nonideal society, they can serve as a tool of oppression. In a sexist and patriarchal society, there is tremendous oppression against the human body, primarily the female body, but not only that, any other body too that is perceived as nonnormative, such as black or nonbinary (Laing 2021) or overweight (Bordo 1993). In such a nonideal society, any increase in monitoring through digital health can raise the degree of control and consequently oppression. Men are also a specifically stigmatized group, against whom function expectations often beyond the capabilities of many men, regarding the realization of the ideals of masculinity understood as strength, poise, perfectionism, and rationality. Men can be both excluded and stigmatized if they do not meet these ideals, or forced to conform to them by biomedical means as well. Patriarchy and sexism are harmful to men as well (Plank 2019; Dembroff 2024).

Feminist Critique of Radical Human Enhancement in Future Space Missions

Feminist criticism of the idea of human enhancement for space exploration draws attention to several elements (Szocik 2023b, 2023d). One is to limit the accessibility of the cosmos by creating a requirement to undergo modification (Schwartz 2020). However, this is context dependent. Modifications can be mandatory or recommended. They may have no negative effects, but they may also have unforeseen consequences. These consequences can occur in space or hinder readaptation to the Earth environment after a space mission. The aforementioned social context also plays an important role. It is not entirely clear whether the economic arguments typical of feminism apply to the space mission environment.

Modifications, even if costly, can be financed by mission organizers or will be applied only to the already wealthy participants in these missions. The financial context works differently here than on Earth, where the financial barrier is cited as one of the main counterarguments. It can therefore be assumed that in the case of the first space missions requiring or suggesting the application of human enhancement, the procedure will not raise ethical controversies. The problem may arise further in the future if space missions were to become bigger and more accessible. Should we put up barriers so that participation in space travel—for example, for work, tourism, or settlement— would require significant interference with our body or psyche (in the case of hypothetical biomodification of our psyche, cognition, or morality)?

The social and political contexts are important here. The question of the legitimacy of such a mission becomes an important issue. Do space missions that require as a prerequisite undergoing radical biomodification aim to benefit the individual participating in the mission, or merely exploit her for the purposes of mission organizers? If the individual will merely be a tool to further the goal of a private agency or the military and political goals of a particular spacefaring state, the moral justification for undergoing such a procedure is negligible.

The rationale for a space mission requiring human enhancement is crucial. If it were to turn out that the justification was in the nature of exploiting mission participants for some privileged entities, then it would become prudent to either postpone the mission or focus on developing alternatives to human enhancement countermeasures. After all, the inefficiency of current countermeasures is the main argument in favor of applying human enhancements, assuming that the application of gene editing, for example, will be easier, safer, and relatively inexpensive.

Feminist bioethicists, in opposing human enhancement and therapy by gene editing, draw attention not only to the status of women and the risks of controlling them, but also to the context of social inequality (Cavaliere 2018). Examples of such social risks associated with both human enhancement and therapy by gene editing include the issue of investing limited financial resources in such technologies, as well as that of equal access. These arguments are legitimate in relation to the terrestrial context, where genetic parenting and biomedical technologies are valued (Cavaliere 2018). These objections do not seem to apply to the space mission environment, where keeping astronauts alive and healthy and, in general, the very concept of human missions consume enormous financial resources compared to the budget of uncrewed missions. Nor would social exclusion be a problem, since gene editing considered in the context of space missions would probably have the status of a mandatory procedure for all mission participants. Thus, while human enhancement applied to space missions would increase the chances of survival for their participants and would be applied to everyone, this does not change the fact that the space mission itself may be organized for reasons that contradict the idea of social justice and, as such, have questionable moral justification from the perspective of feminist ethics.

Consent to apply radical human enhancement for space missions, such as through gene editing, must have strong justifications such as the necessity of protecting health and life, the possible reversibility of the applied procedure, or the lower efficiency of existing alternatives. While human enhancement by biomedical means on Earth is often seen as a kind of superfluous extravagance, this objection loses its force with regard to the space mission environment (Kendal 2020; Szocik 2023a). In space, human enhancement achieves the status of prevention and therapy. The lack of alternative therapies in space that may be available on Earth also plays an important role in justifying human enhancement (Kendal 2022).

Human enhancement serves not only to protect health but at the same time to improve performance, which does not change the fact that certain types of enhancements may be directly aimed at improving performance as performance, rather than as an effect of good health and well-being. Since the environment of space missions resembles that of a battlefield in many respects, and at the same time space policy itself is highly militarized and will probably continue to be so in the future, it is possible to view the concept of human enhancement for future space explorers in a manner appropriate to the military. As Andrew Bickford (2020, 78) shows, military performance enhancements are designed to protect the interests of the state and its investment in training soldiers, not to protect soldiers as individuals for their own sake. The militarization of space exploration will involve the use of military medicine and a military understanding of performance enhancements, with the consequence that astronauts will serve the interests of the state and private space corporations.

The use of genomics in the army could lead to the marginalization of the other components important in preparing a soldier, as well as being used to discriminate in the recruitment process (Bickford 2020, 200). Similarly, the fact that human enhancement is being used in space exploration can lead to the marginalization of the importance of other characteristics, both physical and moral, psychological and behavioral.

On the other hand, the very idea of human enhancement considered for space exploration fits into the utilitarian paradigm that characterizes thinking about space exploration and settlement (Schwartz 2020, 201). Johnson-Schwartz rightly argues that focusing on the idea of human enhancement treated as a necessary element of space exploration, and especially space settlement, can lead to the instrumentalization of humans, who become a tool for an end, rather than a goal. After all, the goal here is space settlement by any means, with instrumental treatment of humans subjected to hypothetical compulsory modification. There is no doubt that Johnson-Schwartz is right about the moral assessment of the risk of instrumentalization of participants in space missions. However, the very idea of human enhancement should be seriously considered as a means of facilitating, and perhaps enabling, adaptation to the conditions of the space environment. Nevertheless, it is worth bearing in mind the critique of the idea of human enhancement offered by Johnson-Schwartz, who also makes its validity dependent on the purpose of, and justification for, space missions, as well as its timing (Schwartz 2020).

It is this pragmatism in the approach to thinking about space exploration that is required here to balance the idealistic approach. The position of feminist bioethics is just such an idealistic perspective. According to Smith and Hylkema (2020, 220), future space exploration at the forefront of settlement will require a departure from the ideals of equality and inclusivity that have not been achieved for many years on Earth. According to the authors, the advantage of the space environment will be that if human enhancement is mandatory for all participants, the problem of equal access will disappear and the charge of social inequality made in bioethical discussions with regard to terrestrial applications will lose justification (Smith and Hylkema 2020, 226).

Feminist Bioethics of Germline Gene Editing

Feminism and Reproduction

The issue of germline gene editing (GGE) is always about reproduction, so it is worth noting possible feminist approaches to the issue of human reproduction in general, for now, regardless of the applicability of GGE. The feminist view of reproduction is not uniform. One of the most extreme differences of opinion on reproduction characterizes radical libertarian feminism and radical cultural feminism. The former views women's reproductive biology as one of the main causes of their historical enslavement and domination by men (Millett 1970). If reproduction were not linked to women, it would be difficult to introduce, maintain, and justify the discrimination against women and their subordination to men. Therefore, radical libertarian feminism considers as an ideal the possibility of making reproduction independent of female biology. One such ideal is artificial reproduction, such as an artificial uterus. As long as only a woman can bear children, a sexual class system persists in which the woman is ultimately always in an inferior, losing position.

Interestingly, some feminists regard the cult of biological reproduction as the source of moral evil in society, which involves the possession of wealth and its inheritance. They speak here of a kind of obsession with possession, the drive to accumulate wealth and then pass it on to biological offspring. If it were not for biological reproduction, there would be no basis for the pursuit of possession and accumulation of wealth and favoritism toward one's own genetically related offspring (Tong 2009, 75). Reproductive biology, understood in this way, through the production of kinship ties, is the source of a tremendous amount of evil in society and politics. It is worth remembering that the cult of biological reproduction goes hand in hand with emphasizing heteronormativity as the only standard (Beck 2021). This model therefore excludes sexually nonbinary people and, through its frequent reference to the family as the basic unit of reproduction, further discriminates against nonbinary people, who, such as in Poland, cannot start a family unless they are a man and a woman. Radical cultural feminism looks at the role of biological reproduction in women's lives and status differently from radical libertarian feminism. Radical cultural feminism sees female reproductive biology as women's greatest asset and their weapon in their struggle against men for liberation, or at least to maintain any meaning in patriarchal society. It holds that the ability to bear children is the one thing that men cannot take away from women or replace, which enforces a certain type of subordination of men to women, as well as their dependence on women, and gives women a certain amount of power and advantage over men (Tong 2009, 77).

Since, from the viewpoint of radical cultural feminism, biological reproduction is a woman's asset, artificial reproduction is, contrary to the view of radical libertarian feminism, a threat to a woman's position. Radical cultural feminism takes a critical view of all forms of reproduction—including in vitro fertilization and surrogacy—that lead to the interruption of the continuity of biological reproduction that characterizes the natural process of biological reproduction, which takes place all the time in the body of one and the same woman. Radical cultural feminism recognizes artificial reproduction as a male effort to alienate women from the process of reproduction so that women lose power and control over it, becoming equal to men, who by nature remain alienated from biological reproduction (Tong 2009, 80). Feminist critiques of reproductive technologies also point to the dangerous and harmful component of experimenting on women, using their bodies as research objects, often ignoring and downplaying the dangers. It is worth adding that the feminist critique of reproductive technologies is not a critique of technologies per se but rather of the risk of their use in power structures that are unjust and unequal to women to reinforce oppression and control over women (LeMoncheck 2002, 150, 154–155).

Feminist reproductive ethics can support or oppose reproductive technologies depending on the circumstances. However, what should guide ethical considerations is the focus on a woman's well-being and interests, as well as her own view of, and ideas about, reproduction, rather than the viewpoint/ good of society or the family, as in nonfeminist ethics (LeMoncheck 2002, 160). For whether feminism supports or opposes ARTs, it does so by recognizing the advantages and disadvantages of these technologies for women (Baylis 2017; Baylis et al. 2020).

An example of the contextualization of the status of reproductive technologies in feminist bioethics is the difference between the status of women and understandings of reproduction and reproductive rights between developed and developing countries. In the case of the former, reproductive technologies can serve the welfare of women. In developing countries such as India, however, reproductive technologies can be used to further increase oppression against women and manipulate them, and can be another tool in the hands of men with power over women. Therefore, as K. Shanthi argues, more important than abstract rights—including reproductive rights—is the improvement of the material and social situation of women in terms of, among other things, opportunities to work, removal of poverty, and elimination of inequality (2004, 119–120). That is why feminists contextualize the discussion of reproductive rights, including the right to abortion, criticizing, for example, the dominance of the ideal of ableism that excludes people with disabilities or other than cisgender (Kendall 2020).

Radical cultural feminism's exposition of women's reproductive biology as their main asset and value, while simultaneously denying reproductive technologies, may look somewhat paradoxical in light of the stereotypical, oppressive image of women in patriarchal society as destined to bear children. As critics of cultural feminism point out, its weakness is that it exposes those feminine qualities that have traditionally been developed under the conditions of an oppressive and patriarchal society (Alcoff 2006, 139). Thus, even if we agree that traits such as caring, kindness, and nurturance are indeed, for various reasons, the domain of women rather than men and, in line with cultural feminism, agree that they are worth nurturing and developing, it is nonetheless worth remembering their at least partially oppressive origins.

This is what the feminist critique of the identification of gender, or the determination of gender by biological sex, is based on. This sense of paradox is reinforced by the aforementioned radical libertarian feminist critique of biological reproduction. Both approaches can be said to have a certain degree of validity, as they both look at social, political, and cultural structures in a different way. The approach of radical libertarian feminism seeks to change the current patriarchal social order rooted in the identification of sex with gender, or in the specific understanding of female sex as destined to deal culturally and socially with what is related to female reproductive biology. In contrast, radical cultural feminism seems more pragmatic in the sense of abandoning the revolutionary ideal of changing the existing order and seeking to improve the role of women within existing patriarchal structures. This does not mean accepting them but interpreting the situation with common sense from the position of the weaker negotiating and political party. In this approach, depriving women of the exclusive right to reproduce, even if only partially, will be symbolic and will lead to depriving women of the only element that marks their uniqueness and at the same time contributes to their oppression and subordination, while simultaneously constituting their only source of strength, namely their reproductive biology.

Socialist feminism recognizes the unidirectionality of both views. The weakness of radical libertarian feminism is its assumption about the conditions of freedom, choice, and informed consent of women, which ignores the actual limitations of the fields of possibilities currently possessed by individual women. The drawback of radical cultural feminism is its essentialist conception of men and women as different ontological beings. Both approaches offer an ahistorical view of sexuality and gender (Tong 2009). Nonetheless, both approaches express valuable intuitions and expose threats, as well as recognizing opportunities for certain solutions.

In a sense, feminist bioethics, as Carolyn McLeod suggests, should maintain a distance from, and caution against, overemphasizing reproductive rights and reproductive autonomy. Since the reproductive function is traditionally associated with women and in many societies reinforces an oppressive and marginalizing image of women seen only in terms of mothers bearing children, the prominence of reproductive rights—in itself legitimate—may as a side effect reinforce an oppressive patriarchy as well as a heteronormative model of social relations (McLeod 2022).

The feminist assessment of reproduction and related issues, including the assessment of the use of ARTs, is therefore complex and ambiguous. The potential benefits of ARTs may be counterbalanced by at least two other phenomena. One is the risk of reinforcing the idea of pronatalism,[5] caused by the possibility that creating reproductive opportunities for previously infertile women or those with fertility problems may reinforce the image of women as obligated to reproduce themselves, this time through technology (Scott 2022, 465). This is a risk that is formally justified (that is, such thinking is formally correct) but, in practice, not encountered in many societies (and therefore the objection is not materially correct). In other words, societies that are conservative enough to judge a woman primarily in reproductive terms are, at the same time, also conservative in their worldview; they are often religious societies, or societies with significant religious influence, which usually oppose the use of ARTs.

Second, the potential benefits associated with the use of ARTs are minimized by the fact that women are burdened by their use, which is also more or less experimental in nature. This is both a physical and psychological burden, where the woman bears the responsibility for the abortion or lack thereof, as well as for the decision to use or refuse ARTs and prenatal diagnosis (PND) (Simonstein 2019). On the other hand, it is difficult for feminism, despite the aforementioned ambiguity caused by the social context and the presence of sexist ideas and structures, to accept a total ban on both ARTs and PND and, even more so, on abortion where ARTs are unavailable or impossible. As Rosamund Scott argues, it is usually women who are responsible for the care of children born with disabilities. In this light, the possibility of ARTs, PND, as well as abortion, is part of the right not only of the woman but also of her partner/family to decide the shape of that family and the way it functions on a daily basis, which the family, and especially the woman, should have the right to choose (Scott 2022, 466–467).

Control and Monitoring of the Female Body in Space: Reproductive Oppression

As we can see, a woman in a sexist society is often viewed through the lens of her reproductive biology and, more broadly, her corporeality. The female body has been a traditional focus of feminism since at least Beauvoir's time. This interest is primarily concerned with oppression and expectations, as well as limitations. A special place is given to the pregnant woman's body, which is a consequence of viewing women through the prism of their reproductive functions. Feminist bioethics, in particular, analyzes the range of challenges and limitations, but also demands, that pregnant women face. The categories dominating this discourse are control, monitoring, surveillance, and responsibility (Corea 1988; Balsamo 1996). Even for women who are not pregnant but are of reproductive age, certain restrictions are imposed, or at the very least, these women face social expectations that have grown up around the idea of the mother. Men of reproductive age are unlikely to be viewed in terms of potential fathers; nor does society place demands on them as it does on women. This fact that men are free from analogous social pressures as well as monitoring despite growing knowledge of the possible significant impact of the quality of the father-to-be's health on the physical condition and health of the child suggests that the monitoring of women and men as potential parents has social, patriarchal, and not necessarily medical roots. As Beauvoir argues, childrearing trends prepare girls for the role of future mothers, and indeed girls are engaged in many of the tasks and responsibilities inherent to mothers regarding childcare. Boys are exempt from being apprenticed to the role of future fathers; in fact, at no stage of childhood are they prepared for it (Beauvoir 2011). For some, and in certain contexts, this could arguably function as an injustice.

Of particular interest to feminist bioethicists is the problem of medicalization. The female body in particular is subjected to medicalization, primarily during pregnancy and in the context of childbirth. Medicalization of pregnant women has negative consequences for women. It leads to imagining pregnancy and childbirth as dangerous phenomena, which consequently require constant medical monitoring. Medical personnel acquire a dominant position over women, who can only passively obey their orders and are subjected to constant control and monitoring (Kukla and Wayne 2018).

If reproduction in space is possible, a bioethically relevant issue is the medicalization status of women participating in space missions. Given the degree of medicalization of pregnant women on Earth, as well as the particularly dangerous space environment, there is a strong rationale for assuming that the degree of medicalization of women will increase, not decrease. Nowadays on Earth, a growing problem of techno-maternity care is identified, instead of a relational, women-centered approach (Sturgill et al. 2019). The space environment may therefore justify particularly increased medicalization and monitoring of the female body, compounding rather than minimizing the oppression of women. Anticipating the risk of enhanced control over women's bodies in space, as well as the possibility of increased medicalization, it is worth bearing in mind the problem of paternalism known from the history of medicine on Earth, which particularly affected women and also overlapped with stereotypical ideas about, and social expectations of, women (Sherwin 1992).

Reproduction in space itself may involve imposing special obligations on women and restrict their rights and autonomy. As long as reproduction in space continues to be a medical and technical challenge, as well as being implemented under an understanding of the space base/habitat as the only or main habitat for humanity, third-party monitoring of, and influence over, women's decisions may gain an additional species dimension. As Beauvoir pointed out, woman was seen as a species-serving creature. An analogous sacrifice may again be expected of a woman. Any discourse on the importance of population growth, as well as political and social campaigns designed to increase population growth, rely on women and seem to implicitly assume their passive consent. The cosmos may be a place where women will be limited in their choices, and the social and legal system may prefer women who choose to reproduce. In a worst-case scenario, women may be forced to reproduce in space in the name of humanity's survival. Alternatively, this oppression will be indirect, and only those women who commit to reproducing will be allowed on missions.[6]

Ectogenesis, or artificial gestation outside a woman's womb, may solve this problem for reproduction in space. In such a situation, women's participation in long-term missions would not depend on their procreative decisions. Regardless of whether or not a female candidate for such space missions would want to have a child, or would have no opinion—and the latter option seems the most realistic, given the long journey in dangerous conditions and the unpredictable individual enduring the physical and psychological challenges of a space mission—human reproduction in space, assuming it were necessary, would be guaranteed regardless of women's consent to their direct participation in reproduction. However, this does not change the fact that, as some feminists rightly argue, ectogenesis can guarantee women's exemption from gestation and childbearing, but no longer necessarily from other traditional jobs assigned to women in patriarchal society, such as housework and childrearing (Cavaliere 2020a; Firestone 1971). In any case, in this particular case of space settlement and reproduction, since the risk of selection based on the declaration of getting pregnant or not only applies to women, ectogenesis can be a solution both for those women who do not want to get pregnant and for those who allow this option but fear pregnancy in space.

The medicalization of women in space in the context of their reproductive functions can be very intense, being a product of a dangerous environment and high technological advancement,[7] as well as perhaps a special concern for saving the species. ARTs used on Earth for years have shown that women inevitably become passive participants in medical procedures supervised by medical experts, usually men. As in the case of in vitro fertilization, the woman's role is merely to lend her body, while the entire process is managed by a doctor through technology (Sherwin 1992, 127–128). It cannot be prejudged that reproduction in space necessarily reinforces these tendencies. However, taking into account the aforementioned factors specific to future space missions, as well as the specifics of the increasing medicalization caused by reproductive technologies, the radicalization of medicalization in space, at least with regard to human reproduction, is very likely.

Is humanity in the cosmos doomed inevitably to replicate the mechanisms of oppression and subjugation of women understood through the prism of their reproductive functions? What about reproductive freedom in future space settlement? Reproductive freedom is supposed to protect procreators' right to free choice and reproductive autonomy from the influence of third parties. The problem arises when procreators' actions affect the welfare of third parties by producing new people (Cavaliere 2020b, 132–133). Giulia Cavaliere (2020b, 134) reminds us that eugenics policies aimed at shaping the structure and size of the population have often been, and still are, of interest to states.

There are strong reasons for supposing that in the future space habitat, which for various reasons will be treated as a new and perhaps the only place of existence for humanity, eugenic policies will be inevitable. Such policies will probably regulate two parameters familiar from the history of eugenics on Earth, namely the size of the population and its structure with respect to desirable or undesirable traits. With regard to the former parameter, technological development will determine the extent of freedom of future space settlers. Their ability to move freely and freely choose where to live in space will be very limited due to their dependence on the life support system. In such a situation, the space base will have limited potential to support the number of people, restricted by resources, habitable spaces, and other factors. This infrastructural element will probably be treated as a factor justifying the abolition of reproductive freedom, understood as the right to decide when to reproduce and how many children to have. Even if such a restriction is justified by infrastructural limitations, it will be controversial because, presumably, some among the inhabitants of the space base will retain the right to reproduce, while the rest will remain excluded from it. Demanding environmental conditions can be taken as justification for unfair reproductive policies that will favor the reproduction of selected individuals. Selection by random lottery is a fair solution, and the above-average health and fitness required for space missions should eliminate any attempt at eugenic policies. However, this does not change the fact that due to the extremely harsh environmental conditions and the aforementioned restrictions on migration options, space, and resources, future space settlement may see eugenic regulations that favor a certain type of trait in future children born on a space base. The application of human enhancement, which will probably be favored for environmental reasons, can equalize the health and performance of all individuals of reproductive age in space and avoid eugenic policies.

Restricting reproductive freedom for eugenic motives is more controversial than restricting it due to the size of the global population. This second type of population policy is justified in terms of the available, very limited space, which physically cannot be expanded, while it too is controversial insofar as space would be an attractive place to live and the number of people willing to make an interplanetary migration, for example from Earth to Mars, would be greater than the number of places. In contrast, any eugenics policy is controversial, discriminatory, and should not be allowed. The solution may be to postpone such a mission until such advances in space technology and engineering make it possible for all interested people to settle in space. Familiar from the history of Western countries of the early twentieth century, the policy of positive and negative eugenics should have no place in a space base. However, it is worth asking whether there is a risk that the specific conditions of the space environment could justify the implementation of such a policy? From the point of view of feminist bioethics sensitive to equality and exclusion, such a policy cannot find justification in the alleged environmental conditions. If, however, it were to turn out that certain health and performance minimums are still required, just as they are required for space missions carried out today, then human enhancement should be considered—assuming it is effective—as a means of abolishing disparities.

The macho cult and ideal of the perfect astronaut that characterize modern space policy culture are leading to a situation in which many parties will expect astronauts, including future space settlers, to perform at a high level. This expectation, in the case of being able to reproduce in a permanent space settlement, will create pressure to produce healthy children who will be able to integrate into the work structures and high-performance requirements inherent in a space habitat (Nesvold 2023, 174). Such environmental conditions, but also the social structure, will lead to increased monitoring of the female body in general and during pregnancy in particular. This particular medical monitoring will not be driven primarily by concern for the well-being of the fetus, but primarily by concern for the well-being of the community, which may make the value of the individual contingent on her usefulness as a worker. One may fear a particularly strong blaming of the pregnant woman for any imperfections in the future child.

There is no doubt that the pro-ableist orientation will be strong in reproductive politics in space settlement. Environmental justifications specific to life in space, as well as social ones, will be added to the motivations already in place on Earth. Environmental arguments will point to inferior adaptation to the harsh conditions of altered gravity, the impact of cosmic radiation, as well as living in isolation. Social arguments will point to the lack of resources and infrastructure needed to care for people with impairments. Such ableist policies lead to negative consequences for people with disabilities.

More pressure would also be put on those women who would not want to reproduce, and the space base would require a certain minimum number of residents to settle. Such hypothetical reproductive pressure would be demanding not only for women but also for nonheteronormative people (Nesvold 2023, 183). In space, anyone can become a victim of reproductive oppression, because not everyone always wants to reproduce even with a fundamentally positive view of the possibility of reproduction. The space environment can also bring many factors that can significantly influence the decision and willingness to reproduce.

Feminist Argument for Germline Gene Editing

Feminist bioethics generally opposes genetic modification, including GGE, for several reasons. One is to emphasize the similarity to eugenics, which is particularly emphasized by feminist disability studies (Labude et al. 2022). Another reason is that GGE expresses the agenda of a perfect society and the ideal of a perfect body, which precludes any deviation from this perfectionist vision (de Melo-Martín 2022a). The final reason is to reinforce the stereotypical role of women as beings destined to give birth, because, after all, GGE is supposed to streamline and improve this process (Nisha 2021; de

Melo-Martín 2023). These are only some of the numerous objections raised by feminist bioethics.

But genetic modification in general, and GGE in particular, need not be opposed to feminist ideals. As Purdy suggests, one can identify some good reasons why prenatal genetic service can be justified from a feminist perspective. One rationale is that there are disabilities that diminish quality of life, and one can imagine a scenario in which a person with these disabilities might prefer not to be born in order to avoid permanent suffering or discomfort. Another rationale is that people with disabilities whose birth was not prevented in the name of the feminist idea of affirming every type of existence require particularly intensive care. This care is usually provided by women, which leads to an additional hardship for females already burdened with responsibilities, including the duty of care. This poses a particularly big problem for women from the most excluded and poorest groups, who do not have the means to pay for special care (Purdy 1996, 82–86). Other feminists respond, not unreasonably, that unjust social relations should be changed (Fourie 2022). Similar counterarguments apply to understanding pregnancy in terms of disempowerment.

The concept of gene editing, like the knowledge of our genetics itself, can increase our knowledge of our bodies. Knowledge, in turn, can raise our control over our bodies and our lives, and thus also give us more options. Both choice and control are overarching feminist categories. Knowledge about our genetics, such as genetic diseases, and the possibilities for treating them, therefore, expand the horizon of choice and control. The assumption that genetic knowledge and gene-editing capabilities will increase a woman's range of choice and control over her own body and life is a postulate that can be realized under ideal conditions. Under the conditions of a society that is oppressive toward women, genetics may increase the control of men and other parties over women because of the perception of women through the lens of their reproductive functions. In this view, genetics will force women to undergo genetic examinations and tests and then make decisions based on genetic knowledge even without their consent (Asch and Geller 1996, 325).

The reproductive context of applied genetics reveals at the same time the role played by the categories of standpoint and particularity. Women's reproductive experience is unique. The male point of view on reproductive genetics often sees only the advantages, that is, the possibility of eliminating diseases and increasing the range of possibilities mentioned. The male point of view is all about increasing freedom, as expressed by Christopher E. Mason's concept of cellular liberty in relation to gene editing (Mason 2021). The male point of view on reproductive genetics is unable to perceive by itself the risk of increasing oppression, especially in those social conditions where women already experience oppression and are treated as a fetal container, even if it is sublimated and not direct.

Feminist Argument against Germline Gene Editing

Feminism critiques GGE, as well as providing grounds for doing so, from a variety of angles. One of them is the interpretation of reproductive technologies, mentioned in the previous section, as a tool of male domination and control over the reproductive process that takes place in the female body. GGE, along with embryo selection, can be seen as a kind of establishment of male control and domination in a process hitherto reserved exclusively for women. Even if, in spite of GGE and embryo selection, pregnancy continues in the woman's body and it is the woman and not the man who gives birth to the child, the man nevertheless obtains a new form of interference and, in a certain sense, even an interruption of the continuity of the process of biological reproduction. Moreover, this interference involves the possibility of controlling and designing his own male vision of the future child and, consequently, the genetic and social profile of the whole society. In patriarchal societies, it is likely to be men rather than women who will decide the nature and extent of the changes implemented in GGE or the criteria for embryo selection. This is another arena in which feminism can criticize GGE, but one that is directly related to, and a consequence of, the first element discussed, the introduction of male participation and control in a hitherto uniform biological process exclusive to women. With the help of GGE and embryo selection, men have a unique opportunity to shape the final "product" of women's reproductive work, that is, the features of the child or the child that is to be born as a result of embryo selection.

Men in patriarchal society are not necessarily interested in the biological realization of the process of reproduction. Men are accustomed to the fact that the biological process of reproduction takes place in the body of a woman. What the man cares about—taking the perspective of a patriarchal system based on repression and domination over women—is the possibility of control.[8] Thus, from this perspective, a woman can continue to function as a contractor, a worker carrying out the difficult and demanding—both biologically and psychologically—process of reproduction. The man, on the other hand, through GGE and other reproductive technologies, becomes the manager of this process.

Feminist bioethics emphasizes that medical technologies are not neutral. With regard to the new reproductive technologies of the 1980s, feminists have argued that they were created by the "white medical and scientific establishment" ("technodocs") to control women. These technologies illustrate the medicalization of female reproductivity. Depending on the circumstances and the target group of women, either female infertility (reproductive technologies for infertile privileged women) or female fertility (birth control for poor women) is given disease status (the concept of infertility as disease versus fertility as disease) (Arditti et al. 1989, xi–xii). The development of reproductive technologies is sometimes seen by feminists as an attempt by men to gain control in the one area where women have a natural advantage over men, namely reproduction. As Genoveffa Corea argues, individual reproductive technologies are taking the place of women at various stages of the reproductive process, while at the same time reinforcing the importance and role, and above all the control, of men in regard to reproduction (Corea 1989, 45). Men's control over women with respect to reproduction stems from viewing women in terms of "reproductive bodies" that serve the purposes of the species rather than the women themselves (Murphy 1989, 68).

The ethical assessment of GGE from the perspective of feminist bioethics is ambiguous. On the one hand, GGE can be considered an integral part of the right to reproduce, especially in those situations where it is the only option for having genetically related offspring free of genetic diseases. The prohibition of GGE thus interferes with the right to reproduce as long as effective reproduction is impossible without the use of GGE. The negative reproduction right therefore means that no party can interfere with someone's reproduction (McLeod 2022, 453).

On the other hand, the right to reproduction, of which GGE may be a part, may stigmatize and discriminate against women as being stereotypically associated with reproduction and childbearing (Brown 2004). Women can be, and can feel, forced, even in a non-direct way, to both reproduce and to refrain from reproducing (McLeod 2022, 455). In this context, GGE proves to be a double-edged weapon. For some women, it provides opportunities to have offspring, while others may be directly or indirectly, or at least feel, coerced into having children, which would be impossible without GGE. GGE can further the sexist paradigm of medicalization of reproduction and be used as a tool to combat the "disease" that reproductive difficulties are considered to be in this paradigm. Any future scenario involving enhancement of the human germline genome will impose obligations on women rather than men, including the risk of creating pressure and perhaps a moral (and perhaps legal) obligation on women to submit to reproductive technologies for the future good of humanity (Simonstein 2019). A future human habitat in space may be a place where space base organizers, citing harsh environmental conditions and the need to protect the survival of the space population, perhaps identified with the entire human species, can treat women with GGE as a means of reproduction.

Feminist bioethics also provides other important arguments against GGE. This is an argument referring to social justice, because the specifics of the application of GGE, as well as the rationale for the procedure, mean that it will affect a small, rather privileged portion of society. De Melo-Martín suggests that public funds should be diverted to finance pre- and postnatal treatments to improve the quality of life and health of children and mothers from marginalized groups. Subsidizing GGE is supporting the privileged, and because of the opportunity cost, resources allocated to GGE mean fewer resources available for use in other areas (de Melo-Martín 2022b).

Another potential negative consequence of GGE is the reproduction of social injustice. As Robert Sparrow (2022) notes, parents in an unjust, nonideal society will choose to edit such traits to ensure their future children belong to a privileged social class, thereby continuing the unjust state of affairs.[9]

Feminist Approach to Germline Gene Editing in Space

If we take a radical cultural feminism perspective on reproduction in general and reproductive technologies in particular, GGE in space will be particularly dangerous to women's interests and status. It will pose a threat far greater than GGE on Earth. In space, GGE, like embryo selection, if it were to be applied routinely for environmental reasons from the beginning of human reproduction in space or for participation in long-term space missions, would involve a form of social engineering. This engineering would involve designing the genetic and social profile of a future society living in space according to the ideas and criteria of those influencing GGE and embryo selection. If the influence of patriarchal ideas were significant, it would be a masculine rather than feminine vision of society and the individual, presumably promoting values appreciated by men and serving men rather than women. GGE applied in a manner characteristic of the processes and structures of patriarchal society will make it possible to create, from the outset, in a relatively easy manner, because it is driven and accelerated by GGE, a society particularly unfavorable to women.

While currently on Earth the hypothetical introduction of the clinical application of GGE will not lead to massive effects, in space, due to the small size of the space population and its closed nature, the newly created society may be an extremely unfavorable environment for women, an extremely oppressive one.[10]

An additional threat may be the perception of space exploration as an activity reserved for men. Difficult environmental conditions may foster the continued exclusion of women and the belief that space missions are primarily the domain of men. This stereotype may be further reinforced by the great distance from Earth and the necessity of separation from family, an area traditionally reserved for women in the patriarchal social vision.[11] If this understanding of space were preserved, allowing women to participate in long-term space exploration would entail assigning them traditionally patriarchal functions, namely servile and reproductive.

In conclusion, the feminist bioethics of space exploration does not oppose the possibility of human reproduction in space as part of future space settlement, the application of human enhancement, or GGE. However, feminism recognizes the dangers that each of these activities may bring. Their biggest victims due to reproductive biology will be women.

Feminist Bioethics of Moral Bioenhancement

To conclude this chapter, I want to mention a specific type of human enhancement—biomedical moral enhancement. The space mission environment poses challenges to the psychological and behavioral fitness of astronauts (Kanas 2011, 2014; De La Torre et al. 2012; Arone et al. 2021). These challenges also have implications regarding those behaviors that can be morally evaluated in terms of good or bad. Already here we encounter the first challenge regarding the morality of behavior in space, for it is not clear which types of behavior will be considered good and which will be considered bad. In addition to the potential role played by environmental conditions for assessing the morality of behavior in space, another factor that may affect the moral assessment of behavior is the interests of those with power. These actors include not only the astronauts themselves but also space agencies, spacefaring countries, private companies, and, finally, humanity as a whole.

Modifying morality through either pharmacological means or gene editing can also be controversial because of the way human morality works. As Susan Dodds (2021) notes, moral agency is a product of embodiment, our identity over time, our understanding of subjective selfhood. The understanding of moral bioenhancement inherent in mainstream, nonfeminist bioethics expresses an abstract, individualistic understanding of the mind in a Cartesian manner as independent of the body and the relationship to the environment. However, it may be that radical moral bioenhancement involving modifying some aspect of the workings of the mind or emotion will not achieve the expected results due to the lack of reference to the relationality and history of our being in the body emphasized by feminist bioethics.

Future long-term space missions, in addition to the obvious risks to physical health and life, will pose mental health challenges. Currently identified challenges for the future include living in isolation for months at a time, monotony, autonomy understood as the lack of direct support from Earth, how to spend leisure time, and a specific understanding of loneliness, meaning separation from loved ones as well as the planet (Le Roy et al. 2023).

Feminist bioethics as proposed in this book skeptically evaluates the idea of biomedical moral enhancement for space missions. This is for two reasons. The first is the technonationalist nature of space policy and the way we think about space exploration. In such a framework of thought and ideology, any attempt to modify morality may serve the goals of the entities benefiting from space exploration, rather than the individuals participating in it. The second reason is the inferior status of women and other excluded groups. What is traditionally the domain of women is considered inferior in a patriarchal culture and may be the main object of modification (Szocik 2023b). There is also a risk of applying moral biomodifications aimed at encouraging mission participants to behave in accordance with the policies of mission organizers. This could include reproductive behavior, depending on current antinatalist or pronatalist policies in space.

Notes

1. For a detailed discussion on human enhancement, definitions, and arguments for and against, see Szocik 2023a.
2. It is not only feminists who criticize this distinction, as such criticism is also made by many nonfeminist bioethicists, although for nonfeminist reasons (see Szocik 2023a).
3. Sue V. Rosser identifies the existence of "androcentric bias" in clinical research as well as diagnosis and treatment, which has negative consequences for women (Rosser 1994).
4. This mindset is reminiscent of an image of the "fetal astronaut."
5. As many feminists, including Kendal, argue, modern societies are essentially pronatalist, for having a genetically related child is considered the fulfillment of the ideal of womanhood despite pointing out objective risks such as the financial situation or climate change (Kendal 2018, 59).
6. See Kendal's chapter on the idea of reproductive slavery for space settlers (Kendal 2023).
7. The risk of medicalization increasing with technological advances is a good example of the thought expressed by Octavio Alfonso Chon Torres. Chon Torres argues that the success or

failure of humanity in space will be determined not so much by technology but by misman-
agement of resources and poor decision-making (Chon Torres 2022, 64).

8. Since all men now live in patriarchy, however, not all respond in a uniform way.

9. One of the weakest and most vulnerable groups, and perhaps in some ways the weakest
when it comes to human enhancement and GGE in space, is the future children who would
be born there. This is because they cannot express their preferences and thoughts. I am
grateful to Koji Tachibana for this thought.

10. It is noteworthy that in discussing many forms of oppression in space, the collective monograph edited by Charles Cockell almost completely ignores gender issues (Cockell 2022).

11. Perhaps planning for the future in space will require an openness to alternative forms of
having children to the family, such as, for example, surrogacy understood as a form of social organization. Surrogacy would undermine the importance attributed to the nuclear
family today (Lewis 2019).

6

Antinatalism, Environmental Ethics, and Feminism

Introduction: The Problem of "Overpopulation" and Climate Change

The issue of antinatalism appears here in a political rather than a philosophical context.[1] Antinatalism understood philosophically assumes that human life is always full of suffering to the extent that the net sum of suffering exceeds that of pleasure. Consequently, we should not reproduce in order to spare suffering for future humans. In contrast, antinatalism understood politically implies a policy of limiting reproduction because of the negative environmental effects caused by "overpopulation." Some authors, such as Hilary Greaves, among others, argue that the problem is not the emission rate, which increases with population growth, but the cumulative emission (Greaves 2019). Consequently, a reduction in population does not necessarily equate to a reduction in cumulative emissions. Antinatalism understood politically does not pay direct attention to the quality of human life, and is not motivated by a desire to spare future people the suffering associated with existence itself.

Another point has important implications for feminist ethics and bioethics. Population reduction can only be accomplished by decreasing or even temporarily prohibiting reproduction. The highest total fertility rates characterize societies in the Global South, which are the poorest, globally excluded, and experience global injustice. Importantly in the context of climate policy, the Global South can be seen as an unjust victim of environmental destruction and exploitation by the Global North (Resnik 2022). Why must the poorest and excluded suffer such drastic consequences of climate change? The feminist perspective not only takes a skeptical view of the concept of population control and "overpopulation," but also rejects blaming the abstract whole of humanity for climate change, pointing out that responsibility for environmental devastation is not distributed evenly among all people (Ojeda et al. 2020). In this book, I am skeptical of the concept of "overpopulation," invoking the term because of its popularity in nonfeminist thinking about global challenges and climate change. It should be emphasized that feminists

Feminist Bioethics in Space. Konrad Szocik, Oxford University Press. © Oxford University Press 2024.
DOI: 10.1093/9780197691076.003.0007

do not regard population growth as an independent factor causing environmental destruction (Wieczorek 2023).

But regardless of the antinatalism politics under consideration, what interests feminism is the imposing of the greatest costs of climate change on the poorest. It is also pointed out that the negative impacts are felt more, and more often, by women than men. Feminism, therefore, introduces a gendered framework of justice and injustice to the ethical discussion of global justice and injustice (Parr 2021). But regardless of the very idea of antinatalism, environmental issues and climate change are bioethical issues because of their impact on health, well-being, and autonomy (Macpherson 2016). Since the most disadvantaged are already the most excluded and marginalized, ecofeminism, due to the centrality of the category of exclusion and oppression, is particularly predestined to analyze these issues. But even nonfeminist bioethics based on principlism is forced to be particularly concerned about environmental justice, which is not experienced by excluded groups, because environmental exploitation and climate change cause the worst consequences precisely in these marginalized groups (Ray and Cooper 2024).

The issue of "overpopulation" and population is particularly challenging for feminism. On the one hand, something fundamental for feminism, especially when confronted with nonfeminist ethics and bioethics that are often oppressive to women, is the acceptance of the right to women's bodily autonomy and independent reproductive decision-making. An additional context that reinforces this central role played by women's rights is the risk of racism that accompanies discussions of "overpopulation" and birth control. On the other hand, as Jade S. Sasser points out, there is the question of feminism taking a stand on population issues (2018, 149). Feminism generally criticizes populationism, whose central tenet is the concept of "overpopulation," instead pointing to the unequal distribution of wealth and resources by elites, as well as the discriminatory nature of "overpopulation" discourse that targets especially the poorest women (Hendrixson et al. 2020). Environmental feminists criticize all forms of populationism, such as demopopulationism (interfering with human populations to get them to the right size and composition), geopopulationism (interfering with populations pertaining to specific places), and biopopulationism (commercialization and commodification of life) (Bhatia et al. 2020). Feminism's opposition to the philosophy of "overpopulation," population control and family planning stems from the fact that for feminists, climate justice is closely linked to reproductive justice, and the latter is violated in a discourse based on population control (Sasser 2023). This discourse is based on the neo-Malthusian ideology of finite resources threatened by population growth (Ojeda et al. 2020).

Despite this anti-population discourse inherent in feminism, there are some views among feminists that are unacceptable under the ideas of climate and reproductive justice. An example is one paper by Donna Haraway: "The incomprehensible but sober number of around 11 billion will only hold if current worldwide birth rates of human babies remain low; if they rise again, all bets are off" (Haraway 2015). Legitimate concern for other animals cannot justify restricting reproductive rights, as only the poorest women in the parts of the world most affected by climate change will be the victims.

It is worth remembering that regardless of the types of antinatalism distinguished here, with the aforementioned distinction between philosophically understood antinatalism and politically understood antinatalism, contemporary feminist bioethics usually opposes antinatalism. The objection is particularly to discriminatory antinatalism, which targets excluded and marginalized groups (McLeod 2022, 458).

In terms of space settlement, "overpopulation," unlike Earth's current and future situation, is not a problem. Overpopulation in space will be easier to avoid due to the nature of life in space being dependent on a life support system and advanced technology. The chapter devotes more space to the situation on Earth than to that in space for the following reason: the problem of climate change and environmental degradation is a serious challenge for humanity. It is worth asking what place there is for space exploration under such challenging Earth conditions. If the global situation on Earth is so alarming and steadily deteriorating, it is difficult to expect space exploration, inherently expensive and available only to the privileged, to have any chance of making a positive contribution to these global problems or to be a model for solving them itself. From the perspective of global feminist bioethics, it is worth asking about the sensibility of such a project as human expansion in space, taking into account the implications and consequences of this endeavor for people and problems on Earth.

Population Reduction versus Greenhouse Gas Emissions

While the argument on reducing population or at least human reproduction rates is often taken for granted in population ethics as a remedy for global warming, there are good reasons to be skeptical of such a thought pattern.[2] Despite the aim for the increasing population to reach a peak around 2100, the population will begin to decline after that peak (Figure 6.1). The only area with a growing population is sub-Saharan Africa, with an estimated increase

World: Total Population

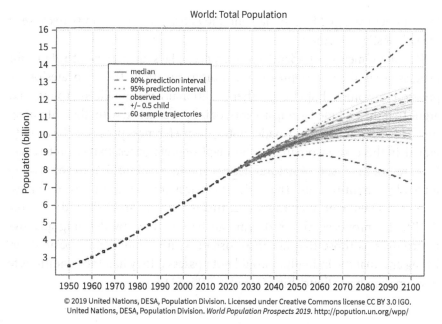

© 2019 United Nations, DESA, Population Division. Licensed under Creative Commons license CC BY 3.0 IGO.
United Nations, DESA, Population Division. *World Population Prospects 2019*. http://popution.un.org/wpp/

Figure 6.1. Estimated global growth forecast for humanity up to 2100. From: United Nations, Department of Economic and Social Affairs Population Dynamics, © 2019 United Nations. Reprinted with the permission of the United Nations (27 June 2023).

from the current one billion to four billion by 2100. Thus, if population policy were really to be about accelerating population decline by, for example, reducing birth rates, sub-Saharan Africa should be the area of implementation of such a policy. This area, however, has a low carbon footprint (Budolfson and Spears 2021). As a result, therefore, the idea of antinatalism, targeting residents of the poorest areas, would not fulfill the function of reducing carbon dioxide emissions, which is assumed by proponents of population reduction. For the problem is the inhabitants of the most developed countries and their lifestyles based on high energy consumption, supported by the military policies of their armies (Ghosh 2021, 174; Szocik and Reiss 2023).[3] However, this does not change the fact that "overpopulation" since at least the 1960s has been treated as a negative phenomenon, and reproductive rights as something that should be curtailed (Hardin 1968; Ehrlich and Ehrlich 2013; Cafaro et al. 2022; O'Sullivan 2023).

The achieved total fertility in the highest-fertility countries coincides with the ideal fertility—the number of children desired by women as optimal— and this number is high (Budolfson and Spears 2021). A woman may in practice not want many children, but she cannot afford to give birth to fewer than the number expected by the man and the patriarchal community because of the risk of stigmatization.

The phenomenon of adaptive preferences plays an important role in the process of eliminating oppression and discrimination. The advocates of oppression of women, but also of other traditionally marginalized groups, may argue that the excluded group accepts, or is even satisfied with, their life situation. However, as Serene J. Khader notes, the preferences possessed by an individual, in this case a woman, may be adaptive preferences, that is, preferences that are not the result of natural desires and inclinations, but in some sense forced by circumstances. These circumstances include centuries of oppression of women, who may think that this is their natural role and position. As Khader proposes, this problem can be identified by defining adaptive preferences as those that are inconsistent with an individual's basic flourishing and that the individual would not possess if they had the conditions to develop their basic flourishing (Khader 2011, 51).[4]

Life Worth Living

A key concept for philosophical antinatalism is the notion of life worth living. Life worth living means the value of life from a personal, rather than a general, point of view. In this sense, the term does not submit to the possibility of an objective determination of its value by a third party but can only be assessed by the individuals concerned, who can recognize their life as a life worth living. An objective evaluation of an individual's life is possible, but it will not be an evaluation in terms of life worth living (Broome 2004, 66–68).

Antinatalism as understood philosophically grows out of a certain type of concern for the good quality of life of future humans, which, as philosophical antinatalism states, can never be guaranteed. Antinatalism is usually shared by people who are relatively well off economically and, at least socioeconomically, are good potential parents. Thus, antinatalism was popular among early feminist thinkers (Firestone 1971; Barrett Meyering 2022). However, it was an attitude that was attractive to better-off women, who had a different reproductive consciousness from poorer women, especially racial minorities, in controlling the reproduction of whom a society driven by racist and, in a sense, eugenic ideas was interested.

Nonetheless, it is better-off people who are more likely to make decisions to postpone procreation in the future because of, among other things, concern for the welfare of future potential offspring. In contrast, those people whose socioeconomic situation is much worse and who, at least in socioeconomic terms, may not be able to guarantee future children a good quality of life are much more likely to reproduce early and have more offspring (they have fast life histories) (Brown and Keefer 2020). From the point of view of the logic of antinatalism, this is paradoxical insofar as we might expect that it is the latter group of people who should be characterized by greater caution and skepticism about their own ability to guarantee parental care as well as the quality of life of future children. However, many reasons drive women to have many children, among them a lack of access to birth control and the need to have children to help support their families.

The concept of life worth living requires the use of some idea of minimum or optimal quality of life. Quality of life is normalized to a certain extent in all circumstances in which third parties are involved in decision-making, for example biomedical decision-making. Feminist bioethics is particularly sensitive to the lives of excluded and marginalized people whose lives may be deemed not to meet the minimum requirements of a normalized conception of quality of life. Judgments about quality of life are subject to at least two types of threats, which are also recognized by nonfeminist bioethics. One is the risk of applying one's own arbitrary preferences to the assessment of an individual's quality of life. The second risk is to analyze an individual's quality of life in terms of its social value (Beauchamp and Childress 2013, 171).

Antinatalism considered as a philosophical idea appears almost inevitably with this kind of consideration, in which we assume that allowing the worst or evil consequences to occur is something irrational, and causing suffering is something immoral. As Matti Häyry argues, these two premises are fulfilled during reproduction because bringing children into the world is worse for them than not existing (the premise of irrationality), as well as causing the creation of another being who will suffer at various stages of life (the premise of immorality) (Häyry 2010, 167–168, 171).

In fact, it is difficult to find a strong and convincing counterargument from a feminist point of view as well. While one could question the absolutism of rationalism and provide counterexamples to suggest the negative effects of the cult of technology and science (assuming that they are indeed the result of the dominance of rationalism), it is difficult to question the second premise, that it is immoral to voluntarily bring preventable suffering into the world. As long as we know that every human being at some stage of her life will suffer, while at the same time, not existing, cannot know and regret what she is losing by never being born and becoming a human being, it is difficult to recognize the supreme value of life and the peculiar obligation to bring more children into the world just for the sake of this vague cult of existence as the highest value. This remark applies only to the decision or counseling to reproduce knowing about the aforementioned consequences in the form of unavoidable suffering, and the only way to prevent it is to refrain from reproduction.

The right to reproductive autonomy has been at the heart of the United Nations (UN) since at least the 1980s. The UN has identified one of the causes of "overpopulation" as the oppression, discrimination, and exclusion of women. In the famous report *Our Common Future*, prepared by the UN World Commission on Environment and Development, the commission negatively diagnoses the impact of population growth on quality of life. According to the report, the critical capacity of individual countries, which were struggling to guarantee education, healthcare, and food security for a growing population, was already being reached (World Commission on Environment and Development 1987, 95). Interestingly, reproductive autonomy, especially for women, is considered, according to the commission, a kind of "basic human right of self-determination." "Overpopulation" makes it difficult to realize human potential (World Commission on Environment and Development 1987, 96). This thinking aligns with Martha Nussbaum's (2000) concept of ten central capabilities and emphasizing the importance of women's development. One of these capabilities is bodily integrity, which means, among other things, the right to autonomy and reproductive freedom (Nussbaum 2011, 33). As Gro Harlem Brundtland, the head of the World Commission on Environment and Development, argues, educating women has far more benefits in terms of social returns than educating men, because educated women have fewer children. They also know how to control their own reproduction and have the required capabilities (Brundtland 1993, 898; 1994, 18–19).

The commission stresses the need to improve health and education specifically for women, as well as all other comprehensive measures to raise the social status of women and ultimately reduce the rate of population growth (World Commission on Environment and Development 1987, 103, 104, 112, 140). The commission points to a clear correlation between increasing employment opportunities for women, access to education, delaying the age at which they marry, and lowering fertility rates. It explicitly mentions the need to promote women's rights and interests (World Commission on Environment and Development 1987, 106, 124–125, 141, 257).

In conclusion, therefore, it can be said that, in light of this approach presented by the UN commission, guaranteeing a life worth living for future generations requires first guaranteeing a life worth living for women, especially those of reproductive age. This idea is undoubtedly feminist in nature.

Feminist Arguments for Antinatalism:
The Case of Ectogenesis

If we assume that in the patriarchal model a woman is perceived through the prism of her present, potential, or past reproductive functions, it can be said that antinatalism, despite its extreme character, at least theoretically and at least in some respects can be an attractive alternative to patriarchal pronatalism. Antinatalism understood in terms of women's struggle for reproductive autonomy is not so much focused on preventing the suffering of future children, as David Benatar, among others, emphasizes in his concept of philanthropic antinatalism, but on protecting women's rights and freedom. Thus understood, philosophical-political antinatalism frees woman from the reproductive function, which determined the way she was perceived in society and subsequently deprived her of many rights and opportunities for development. This feminist understanding of antinatalism dictated by concern for women leads to a paradox caused by the nonfeminist idea of extending the human species, which is done with the involvement of women. Therefore, antinatalism shared by women is perceived negatively in a patriarchal and sexist value system in which women are identified with their reproductive function. The antinatalism considered by feminist bioethicists is not seen as a tool to combat "overpopulation" (McLeod 2022).

Antinatalism can be understood as the radical outcome of reproductive autonomy. As Firestone has pointed out, the first step on the road to women's equality must be the liberation of women from the oppression and constraints imposed on her by her reproductive biology. This liberation should consist in spreading the responsibilities of childbearing and childrearing as far as possible over society as a whole and thus also over men (Firestone 1971, 206).[5] Feminine biology itself places a burden on women—in a sense the result of the unfair lottery of natural selection. The idea of female biological specificity, interpreted as a source of disempowerment and physiological burden, is widely exposed by Beauvoir (2011) and Firestone (1971), among others. Firestone adds that this natural injustice was later exploited and compounded by men who reinforced it in cultural institutions (Firestone 1971, 205). This belief led to the perception of technological advances as a means to end this biologically rooted oppression of pregnant women (Hester 2018, 14).

Biology itself is not necessarily a source of disempowerment. Instead, a woman's biology has become the source of her disempowerment within patriarchal structures. Women's reproductive biology is often used to control women, and both law and custom in many countries restrict women's reproductive rights (Cook 2002, 47).

This is not to say that Firestone's idea of liberating women through reproductive technologies was based on a misdiagnosis of the position of women. It is possible to imagine situations and circumstances in which women's reproductive biology would constitute their disadvantage compared to men even in a nonpatriarchal society. Nevertheless, Firestone failed to explain the genesis of this oppression and disadvantage women are led to by their biology (Oakley 1997, 31). Women's reproductive biology is not necessarily a problem or burden; it becomes one in a particular social environment. This is what Firestone's critique points to.

Some feminist bioethicists today also still see opportunities to reduce women's oppression through new reproductive technologies, such as ectogenesis, which means the possibility of developing an embryo and fetus outside the body of a living organism and requires an artificial uterus. The first task of ectogenesis of interest to feminism is to correct the unequal situation of women caused by their reproductive biology. But we wanted to point out that ectogenesis can also be useful when considering antinatalism. The two do not seem related, since ectogenesis is supposed to enable reproduction, while antinatalism opposes it. But women are always disadvantaged and at a disadvantage with any population policy, whether pronatalist or antinatalist. Ectogenesis, therefore, can reduce the negative image of women that supports the philosophy and policies of antinatalism, because it allows reproduction without the involvement of a specific woman. In many countries, pronatalism is supported by authorities and ideologies that are nationalistic and anti-immigrant, and women are the target of these policies because they are understood as a reproductive tool. Women who are not interested in reproduction, as well as those who actively promote antinatalism, are seen as selfish, as well as not fulfilling the female destiny. Ectogenesis, therefore, enables women to maintain their antinatalist philosophy, as well as relieves them of blame for causing growth under conditions of "overpopulation."

The undoubted advantage of ectogenesis—assuming it would be applicable to the human species in the future—is that it breaks the association of women with the role of mother and reproduction. Such an association has many negative consequences, one of which is pronatalist oppression, according to which infertility is something evaluated negatively and usually burdens the woman (MacKay 2020, 347–348). As Kathryn MacKay rightly argues, since pregnancy is a dangerous process for a woman, getting pregnant must not be the result of third-party pressure, but should be solely the result of the desire of the woman interested in pregnancy. In a situation where ectogenesis would be available, it should be introduced if the alternative is traditional pregnancy associated with many risks and negative consequences for the woman (MacKay

2020, 350). Evie Kendal (2022, 213) adds that ectogenesis is always beneficial for women, because it eliminates the risk of health complications, protects against possible negative consequences of pregnancy on the labor market, and gives equal opportunity to parties other than women to have their own offspring.

Some feminist bioethicists have criticized Mackay's concept by pointing out that abandoning female reproductive biology by replacing pregnancy with ectogenesis is not an appropriate way to abolish gender-based oppression. Giulia Cavaliere notes that women are repressed and under oppression not only and not always because of their identification with their potential reproductive functions, but because of their sexualization and other stereotypically understood characteristics and expectations. The introduction of ectogenesis would be to bring women in line with the ideal of male biology and male reproduction and eliminate what is inherent only in women (Cavaliere 2020a, 729–730). Moreover, in a nonideal patriarchal society, both ectogenesis and other technologies, including reproductive technologies, originally planned to liberate women, can be used to further oppress them. Cavaliere (2020b) adds that ectogenesis would lose its equality- and freedom-promoting potential for women due to unfair social relations that promote women in caregiving roles, as well as privileging men in the labor market.

Ectogenesis may find a particular application in space. In addition to the advantages mentioned above for Earth, space exploration adds several new ones. Kendal argues that ectogenesis would make it possible to transport frozen embryos from Earth to a space base, thus speeding up the process of space settlement, as well as ruling out the risks associated with space pregnancy and excluding pregnant women from space base activities. Kendal (2022, 219) adds that ectogenesis would also solve the problem of crew selection in terms of gender, where one has in mind an appropriate proportion of women of reproductive age with an eye toward space settlement. In a future space exploration scenario involving human reproduction, ectogenesis has the potential, if not to completely eliminate, then at least to minimize the oppression and discrimination traditionally caused by reproductive biology on Earth.

In conclusion, feminist bioethicists argue that reproductive biology need not be inherently a source of disempowerment for women. However, in practice it often is, due to patriarchal social structures. Even today, in Western countries with the highest degree of gender equality and of feminist consciousness, there are disadvantageous inequalities for women derived from their reproductive biology, such as wage inequality. It is also worth noting, as some philosophers point out, that adopting antinatalism as a total philosophy and practice would solve all the problems so central to feminism, namely the degradation of the planet caused by humans. The disappearance of all humans would equal the disappearance of all problems (Räsänen and Häyry 2023).

Feminist Arguments against Antinatalism

However, not for every woman is her biology a source of disempowerment. Because different groups in society have varied interests and problems, the same is also true for differences among women in what distinct social, economic, and ethnic groups see as goals and threats. This has been a familiar problem in feminist philosophy for decades, and its essence may be to draw attention to the monopolization of feminist discourse by white, middle-class feminists from affluent Western societies. This problem was expressed, among other things, by pointing out that what was a problem and a kind of oppression for the white, middle-class woman could remain a dream for her colleagues from developing countries. This difference between interests and distinct perceptions of problems and challenges between different groups of women has important implications for the issue of antinatalism. For white, middle-class women, what mattered most during the second wave of feminism was the containment of pronatalism. Its effect could be both antinatalism in the aforementioned political sense of freeing women from the sole or dominant procreative function, and simply reproductive freedom, not necessarily in the antinatalist sense. Since women are different from each other, not all of them felt the stigma of pronatalism equally. On the opposite side were women who were discouraged from reproducing or from having contact with their own children even if only as a result of their need to provide for their families. Thus, the function of a housewife was not always associated for all women with oppression. What may be good or simply neutral in itself, that is, reproductive freedom and autonomy, which can manifest itself both in temporary or permanent abandonment of procreation and in antinatalism, due to the exclusion and marginalization of poor women in both developing and developed countries, made it difficult for them to realize reproductive freedom in the opposite direction, that is, pronatalism. Difficult economic conditions and structural exclusion make it impossible or difficult to guarantee a "life worth living" for future children and provide arguments for antinatalist policies and ethics. Consequently, as McLeod notes, for many excluded women the problem is not the obligation to reproduce but its de facto prohibition, or at least the many restrictions and obstacles to the realization of biological reproduction (McLeod 2022).

The wide variation in women's social and economic status makes it impossible to take a single bioethical position on women's reproductive rights and reproductive autonomy. Feminist bioethics both supports antinatalism where the woman is seen as a fetal container and defends the right to biological reproduction where women's reproductive rights are taken away or restricted because of the poor quality of life of the mother and the presumed poor quality of life of future children. The problem is a situation in which contraception and abortion are treated not as the result of a woman's genuinely free choice, but as a consequence of environmental pressures and her poor quality of life, which create conditions (structures) that are impossible to transcend. The coercion to undergo an abortion or use contraception, even if indirect, is no different from the coercion or reproductive pressure. The problem is compounded by girls' being brought up under patriarchal conditions, which leads to the evolution of the aforementioned false consciousness, where women identify their interests with those of the social system that exploits them. The strength and extent of both pro- and antinatalist oppressiveness varies and functions differently depending on the identity of individual women, and has an incomparably more negative impact on women of color (Wilson et al. 2019). Intersectional feminists show that particular combinations of different identities can lead to greater vulnerabilities caused by climate change due to the weaker position of women, particularly women of color (Sasser 2023).

It can be said that antinatalism was the natural ally and natural outcome of second-wave feminism, geared toward liberating women from the oppression of pronatalism. But soon after this alliance between feminism and antinatalism, the relationship between them was weakened. One of the reasons was the consideration of the global context, which shows that different social and ethnic groups of women have different problems, and pronatalism in a sense is (or rather was) a specifically Western problem, while the inability to free oneself from antinatalism is a specific problem for excluded and marginalized groups of women. Moreover, poor women of the Global South are now being targeted by antinatalist policies that equate pro-environmentalism and responsibility with delaying or even abstaining from reproduction (Schultz 2010).

Another factor that weakens the connection between feminism and antinatalism is the ethic of care as understood by cultural feminists (Coole 2021), which sees women's reproductive biology as a source of power, rather than disempowerment, as a unique value. This perspective in a sense rediscovers and reevaluates the value of motherhood and parenthood, liberated from patriarchal pronatalism, but also opposed to antinatalism. The feminist response to patriarchal pronatalism does not have to be exclusively antinatalism, but nonpatriarchal, feminist natalism or pronatalism, or simply a celebration of female reproductivity.

Antinatalism as such is not a racist concept. Instead, it can lead to racist consequences when it is not considered *in abstracto*, purely in terms of philosophical speculation in the manner proposed by Benatar and Häyry, but rather in terms of a practical population policy. Since high rates of reproduction now characterize residents of the Global South, both in their home countries and often as migrants living in the Global North, it is inevitable that it is usually residents of the Global South (or migrants from the Global South) who become the focus of antinatalist considerations of a political and social nature. This racism generated by political and social antinatalism may be indirect, when the logical conclusion of antinatalism is that reproduction should be restricted among the inhabitants of the Global South, or it may be direct, when some authors express criticism and dislike of the reproductive practices and lifestyles of the inhabitants of the Global South (see Seymour 2013, 207).

Focusing on "overpopulation" and considering the reduction of reproduction rates as the main goal of climate policy ignores the aforementioned fact that the goal of climate policy should be to reduce cumulative emissions. Such emissions and environmental destruction have many causes unrelated to population numbers, such as colonization, wars, and the accumulation of global resources by the richest. It is therefore a mistake to equate environmental destruction with "overpopulation" (Seymour 2013, 208). The negative environmental impact of one resident of the Global North is incomparably greater than the corresponding impact of one resident of the Global South. From this perspective, antinatalism understood in terms of climate policy has no justification. Antinatalism would be justified only if a high degree of reproduction were possessed by those populations that most pollute the environment. However, this does not apply to the fastest-reproducing populations of the Global South. Consequently, antinatalism may still remain an attractive, abstract philosophical position centered around categories such as happiness, pleasure, and well-being, but it is not a good candidate for the climate change public policy. The thinking that leads to antinatalist conclusions stems from the idea of the Anthropocene that humans have a negative impact on the environment. But what is not properly separated in this popular idea is the frequent equating of the impact of human activity on nature with the number of people (Ojeda et al. 2020).

The problem of "overpopulation" recurs frequently in discussions on climate change. It is usually suggested that family planning and fertility rates should be reduced, which can lead to a reduction in reproductive rights. Feminist bioethics aims to modify this thinking, which places responsibility for climate change on the poorest women of the Global South. This paradigm, by promoting the idea of reducing their reproductive rights, creates their image as the only or main people who can save humanity from climate annihilation (Bee and Park 2022). Interestingly, this paradigm places responsibility on, and limits the rights of, the most marginalized and excluded, instead of emphasizing the need to change consumer habits and, above all, structural solutions to energy consumption. In addition to drawing attention to the global distribution of the "overpopulation" problem and the consequences for the most excluded groups, another alternative is to modify the nuclear family model by reducing the number of offspring through multiparenting (Gheaus 2019).[6] This concept is feminist in nature, as it opposes nondiscriminatory antinatalism, respects reproductive rights, and at the same time includes an element of concern for the environment and limited resources. It also takes into account the idea of intergenerational justice and sustainability.

In conclusion, the problem with antinatalism taken up in population growth discussions is that it has often been treated as a tool to stop population growth. And it has also usually been addressed to poorer, marginalized populations with high birth rates. This has brought the concept under feminist criticism.

Feminism and Antinatalism in the Context of Space Settlement

The expected low quality of life in at least the initial stages of space settlement provides a strong argument for antinatalism and may undermine the concept of saving humanity in the long term through space settlement (Szocik 2023). Some, like Phil Torres, while not pointing to antinatalism as an alternative to space settlement, are equally skeptical about the quality of life for future generations inhabiting a settled cosmos.[7] Torres's main argument against space settlement, or at least in favor of delaying it as far as possible, is that human populations in space will develop into multiple warring groups that will exist in a perpetual state of Hobbesian conflict and mutual fear of each other (Torres 2018).

If reproduction in space is possible (and necessary to guarantee the continued existence of humanity), the model of space settlement organization may replicate the inequalities known on Earth in terms of the limited right to reproduction for certain groups. These groups currently include, but are not limited to, the sexually nonbinary, the disabled, and the poor, who are sterilized, who have their access to infertility remedies restricted, or who have their social assistance restricted and consequently have their rights to raise children taken away (McLeod 2022, 452). It is also a reproductive injustice manifested in higher infant mortality rates for black people regardless of social class (Davis 2019).

Perhaps, however, the peculiarities of space missions will paradoxically make the exclusion of nonwhite women a nonissue. In the case of the struggle for abortion rights in the United States, the interests and rights of white women were different from those of black and Puerto Rican women. The latter were not infrequently forced or encouraged to be sterilized, while white women, in contrast, found it difficult to obtain sterilization or were forced or encouraged to give birth (Cameron 2020, 39–40). Racism in this case was linked to the socioeconomic situation of the excluded nonwhite women. In the case of space exploration, it can be assumed that the socioeconomic status of future participants in long-term missions will be identical, so justifying racist segregation with regard to reproductive rights will be more difficult. Racism will probably no longer be able to be motivated by current economic exclusion. However, it may resemble the situation of the black middle class in the United States, who, even if they match the white middle class in terms of living standards, are far behind them in terms of global capital and wealth accumulation.

However, in addition to this hypothetical advantage of the space mission environment over Earth, space conceals a dangerous potential from the point of view of oppression in the very idea of space settlement understood as space refuge. Assuming that humanity in the future under certain specific circumstances may treat space settlement as the only form of extending the human species, there is a risk that women's reproductive rights will be restricted and subordinated to the overriding idea of saving the species. This could take the form of forcing women to reproduce and making such a promise a condition of being able to participate in a space mission. Given that, despite some moral progress, we still live in a nonideal society that discriminates against others on the basis of gender, class, or race, it can be assumed that these forms of discrimination will play a role in the selection of the crew of a space mission understood as a form of space refuge.[8]

A danger identified by feminist bioethics, but usually overlooked by non-feminist bioethics, is the risk of promoting the idea of compulsory heterosexuality in the name of guaranteeing the survival of our species in the future (Hester 2018, 54). The risk of compulsory heterosexuality may be particularly high in the context of the idea of space exploration understood as a way to save humanity. Both in the context of the issue of human enhancement discussed in the previous chapter and in the concept of antinatalism discussed here in the context of our future in space settlement, heterosexuality is usually taken for granted as the norm. Feminist bioethics draws attention to the potential negative implications of the concept of heterosexuality as the norm (Cameron 2022), with its implications for both non-heterosexuals and the disabled, who may be discriminated against in future space policy due to reproductive politics.

Living More Livable Only for the Rich and Privileged?

In specific cases, the situation on Earth may determine the abandonment of any space policy, or simply make it impossible. Therefore, it is worthwhile, when considering future space exploration scenarios, including the concept of space refuge and space settlement, to pay more attention to the hypothetical conditions of life on Earth in the next hundred years.

We assume that progressive desertification,[9] the sinking of many cities by rising sea levels, and rising temperatures will make life in many parts of the world impossible without technological adaptation.[10] To this list should be added meteorological disasters, which will be frequent and of high intensity (McGuire 2005, 30). By technological adaptation, we mean appropriately advanced technology that may be able to counter these effects but faces at least two constraints: financial and population. The financial constraint means that the countries most affected by these negative climate impacts will not have sufficient financial resources, or perhaps access to the technologies that could enable continued existence in threatened areas. But even overcoming the financial constraint runs up against the population constraint. Providing adequate technological adaptation for the billions of people living in these regions is unlikely to be possible because of the difficulty of continually maintaining advanced technology and producing massive amounts of energy that, paradoxically, will exacerbate climate change.

What will happen to the billions, or at least hundreds of millions, of people living in the poorest but also most vulnerable areas of the Global South due to the negative effects of climate warming?[11] A huge part of them will die because of the lack of access to drinking water, the flooding of cities, a lack of food, and finally high temperatures making it impossible to survive. At least some of the inhabitants of these regions will attempt to migrate. But these migrations, in order to be effective, will require settling in northern countries, which will also be affected by the negative effects of climate change and environmental pollution. It is difficult to suppose that these countries will accept hundreds of millions or even billions of people into their territories.

Is the only fate for the inhabitants of the excluded Global South to be death caused by climate change? This is the worst possible future scenario, but it is not only possible, it is probable. Perhaps this is an unsolvable problem if we recognize that there are technological and population constraints that beyond a certain critical point cannot be solved. But regardless, it would be a situation that poses a serious moral challenge to humanity, because it means that we are doing nothing to improve the lives of the future billions of people in the most affected areas of the Global South.

It is quite possible that people will not be able to live and survive beyond the minimal level of quality of life, which cannot be provided outside shelters due to effects of climate change and environmental pollution. By a minimum level of quality of life, we mean the level that requires the aforementioned advanced technological adaptations that will not be possible to guarantee for hundreds of millions/billions of people in the Global South. Since their quality of life is lower than that of northerners, southerners will not have the resources necessary to enable them to continue to exist under new and extremely harsh environmental conditions.

The richest countries will probably also struggle with climate challenges. Climate wars are inevitable, caused by resource scarcity, inequality, and migration movements (Beard et al. 2021). Although our ancestors experienced major climate changes, these changes occurred much more slowly and Earth's population was much smaller than it is today.

This leads to the admittedly repugnant but paradoxical conclusion that the quality of life a hundred years from now will not be as low as we may think, because the continued reduction in quality of life due to climate change will at some point lead to an inability to continue living. Consequently, life on the planet will be so difficult that any possible life will require a certain minimum level (i.e., a certain minimum of technological adaptations) that will be impossible for the aforementioned hundreds of millions or even a few billion inhabitants of the Global South.[12] The minimum level of well-being of the surviving population will not be maintained by systematically improving the well-being of those below a certain minimum standard of living. We can call this level of life barely worth living or even not worth living.[13] It is not possible to increase the standard of living of an ever-increasing population with constantly deteriorating environmental conditions. It has to get worse. Thus, the average minimum well-being of the global population will be achieved by the likely death of those least adapted to climate change. Not all people will be able to survive, but those who will survive, and then who will be born, will have to have minimal required living conditions provided.[14]

According to some calculations, after reaching about 11 billion people around 2100, the human population could drop as low as 2.3 billion by 2300 (Gowdy 2020). The mitigation strategy, if available at all, will be adopted only by the wealthiest individuals and the wealthiest countries. The basic strategy will become that of adaptation to catastrophic climatic conditions, an attempt to survive on reasonably still habitable land. The poorest people in the Global South will be technologically excluded not only because they lack the means and technology. They will not be supported by a strong state with a powerful army, which may be inevitable in the coming era of climate wars over access to habitable land, drinking water, and farmland. The migration necessary for survival for hundreds of millions of people in the regions most affected by climate change will be held back by the armies of rich countries.

This scenario may seem counterintuitive and contrary to the belief that the future will lead to a worsening quality of life due to "overpopulation" and climate change. Just imagine the effects of a global temperature increase of two degrees over the next few decades and the consequences of this change if only for rising water levels (Jenkins 2021, 178). Increasing the temperature by a few degrees will significantly worsen living conditions. Surviving under climate catastrophe conditions will require expensive, highly specialized, and energy-intensive technology not available to the poorest.

This is a hypothetical future scenario in which humans drive a sizable portion of humanity to extinction (Tyszczuk 2021). This vision is based on the idea of humanity's identity as a being unable to stop the exploitation of the natural environment. The scenario assumed here is one of the worst-case scenarios, for this scenario assumes that the future of humanity is no longer open, and all that remains is to try to adapt to the inevitable climate catastrophe rather than mitigate it (Bowden 2021).

Assuming that such a future is inevitable, we can look at the antinatalism discussed earlier in a slightly different light, just understood more philosophically (avoiding inevitable suffering) than politically (one version of political antinatalism is feminist defense of women's rights; the other is racist targeting of those most reproducing). While the very idea of antinatalism may provoke outrage and conflict with our intuitions in which we rejoice in each new life rather than lament the fact that another new person has come into the world, the reality is that antinatalism lies behind any concept that highlights the problem of planetary "overpopulation," especially in the context of climate change. However, the difference between the philosophical understanding of antinatalism presented in this worst-case scenario and the political and social understanding is as follows. Philosophical antinatalism grows out of a concern for the well-being of future humans, which will be very low. Understood in this way, antinatalism could be seen as a specific kind of care ethics and feminist ethics, as long as the generation making the decision to stop repro- ducing is not interpreted through the lens of paternalism deciding the fate of future people. Reconciling these perspectives is certainly problematic. Antinatalism as discussed here thus arises only as a result of compassion to- ward the expected suffering of future humans and is not understood in terms of population policy for the purposes of climate change.

In contrast to philosophical antinatalism, political antinatalism emphasizes the negative consequences of excess humans on the population as a whole and on the environment rather than on themselves. The very fact of existing under conditions of "overpopulation" and an additionally damaged environment presupposes a reduction in the quality of life of each additional individual born, not just those already in existence. Nevertheless, political antinatalism speaks of humanity and the population as such, pointing to birth reduction as a necessary tool. This is not philosophical antinatalism in the strict sense, for the political justification does not speak of causing harm by bringing new human beings into existence. Political antinatalism shares features with mis- anthropic antinatalism, which states that we should not reproduce because future humans will cause suffering and problems for others. This is the idea behind policies that restrict reproduction, because they want to minimize the problems caused by an excess of people.

The Impact of Climate on Human Development

The ability of humans to survive over thousands of years must be adjusted for the very small total population tens of thousands of years ago compared to the population today. This is a critical element that accounts for the qualitative difference between the adaptive capacity of our ancestors and humans today. It is this difference between us and our ancestors that Benjamin Lieberman and Elizabeth Gordon point to when they emphasize the adaptive capacity of our ancestors consisting mainly of the ability to change food sources and the ability to develop and use even those areas that are not suitable for per- manent habitation (Lieberman and Gordon 2018, 27–28). Under today's conditions, and especially a century later, it is hard to imagine that hundreds of millions of people will be able to quickly switch to alternative ways of pro- ducing food, especially under conditions of a damaged environment and ele- vated temperatures that will not only reduce crop yields but also increase the amount of land that is unsuitable for farming.

In emphasizing the importance of climate to the development of civilizations, we are not just referring to historical examples of the collapse of civilizations caused by climate change (Diamond 2005). We are thinking of John Gowdy's essay in which he suggests that progressive climate change will lead to the collapse of civilizations as a result of making agriculture, which requires a reasonably stable climate, impossible. The consequence of the collapse of civilization is supposed to be a return to hunter-gatherer structures, those preagricultural structures appropriate to the era of climate change in the Pleistocene (Gowdy 2020). Thus, human-induced climate change will cause a return to Pleistocene climate instability. The implementation of agriculture became possible only after the climate stabilized in the Holocene, when the average temperature increased.

As Gowdy concludes, our hunter-gatherer future may be the best possible future amid the inevitable climate catastrophe and subsequent collapse of civilization. The problem mentioned is "overpopulation" of the planet. There are too many people to provide relief, rescue, and a good quality of life for all the billions of an ever-growing population. There are too many people on Earth to allow hundreds of millions, or several billion people, to take up gathering and hunting after the collapse of civilization. Therefore, a scenario in which the inhabitants of the richest countries will try at all costs to maintain the highest possible standard of living and fight the effects of global warming, probably isolating themselves from the poorest, and at the same time most affected by the effects of climate catastrophe, seems quite likely (Jones 2021).

By contrast, regardless of how, after social collapse, humanity organizes itself socially and politically, social collapse itself seems inevitable despite the fact that climate change is linear and slow, which may give the impression of its low potential on the scale of the degree of risk inherent in global catastrophic risks (Frame and Allen 2008, 282). As Robin Hanson argues, today's societies are uniquely susceptible to collapse even with the impact of a seemingly small factor. The main reason for this vulnerability is the constant dependence on enormous amounts of energy and the interdependence of many different systems (Hanson 2008, 366–367). Consequently, that portion of the population that may enjoy a life worth living may lose their current well-being relatively quickly and easily.

Quality of Life and a Life Worth Living for the Entire Population

A life worth living is a life in which moments of pleasure and of no suffering quantitatively outweigh moments of suffering. The absence of suffering is a more desirable state than the state of pleasure. For the sake of the argument, the possibility of satisfying higher needs, which include intellectual and spiritual needs, including science and art, is omitted. For many philosophers, these constitute the essence of humanity, because some of them, such as Polish philosopher Roman Ingarden, believe that to be human is to transcend animalism, that is, extend beyond the satisfaction of biological needs (Ingarden 1983). This utilitarianism included in the concept of a life worth living contradicts a long-standing philosophical tradition expressed by Aristotle in his *Nicomachean Ethics*, according to which no one could be happy with a life full of carefree, sensual pleasures (Aristotle 2002). For this view, a truly valuable life is one that is endowed with the pleasures of our highest capacities. A great life can therefore be full of moments of struggle and sacrifice that lead to truly inspiring achievements. Let us assume, however, that we can speak of a life of minimal good quality when suffering, including psychological suffering, is surpassed by moments of pleasure and a lack of suffering.

The basic prerequisite for well-being, which is the basis of a minimal quality of life, a life worth living, is the satisfaction of bodily needs. Hunger, thirst, disease, a high crime rate, and a lack of shelter from the vexing effects of climate are major factors that drastically reduce the quality of life. A sizable portion of the population, primarily in the Global South, suffers from some or all of these factors that reduce quality of life. These people have a low quality of life.

Quality of life is a state determined subjectively, but only to a certain extent. There are certain circumstances that enhance quality of life. There are also circumstances that almost always decrease quality of life. The lives of a huge portion of the population can be considered barely worth living. Climate change will reduce this already low quality of life to such an extent that the lives of people in the Global South will go from being barely worth living to being life not worth living. It can be assumed that inhabitants of regions affected by droughts, desertification, and flooding will attempt to migrate to parts of Earth where a life worth living can be expected (Jenkins 2021, 193). Desertification affects a quarter of the planet's surface and about 20 percent of the world's population (Beard et al. 2021).

The situation of the poorest around the year 2200, partly as a result of climate change and limited access to water and food, may worsen to the point where, according to some futurologists, future generations may witness the birth of the business of human hunting (Wright 2016). Expanding on the controversial idea proposed by Daniel William Mackenzie Wright regarding manhunting tourism, one can imagine a situation in which the entertainment-hungry rich can offer to help the poor survive on the condition that they participate in the survival game they offer. But regardless of how cultural patterns may develop and, above all, what may happen to moral norms on which there is relative consensus today, it can be assumed that there is a certain population, a certain number of resources, as well as a certain degree of acceptable exploitation and environmental degradation, beyond which there is an irreversible decline in the quality of life for a large portion of the global population.

Space Settlement as an Alternative?

Space settlement is often presented in terms of a means to save the existence of the human species, or at least increase its chances of survival in the event of global catastrophic and existential risks. Space settlement understood literally as intended to guarantee the continued existence of the human species is a reasonable idea. In contrast, space settlement understood as an alternative to solving population and environmental problems on Earth is unrealistic with the time frame discussed here. Space settlement therefore has limited potential to save those most affected by climate change, which does not preclude considering space settlement as a reasonable countermeasure for other types of disasters that may come in the distant future.[15] The very idea of space refuge can be defended in terms of one possible future scenario, which under certain specific conditions could be realistic, and perhaps even desirable. These conditions are the identification of the type of catastrophe from which Earth cannot be protected, as well as the timely planning of a long-term, possibly multigenerational, implementation of such a space refuge. Space refuge is not considered an ad hoc solution. Rather, I understand the concept in a somewhat idealistic way, as a mature decision by humanity to recognize the sensibility of saving at least a small part of the human species.

Feminist space ethics and space bioethics make the case that space exploitation and exploration is even more exclusionary, and reserved exclusively for the dominant, wealthy portion of the population. The space mission environment thus replicates patterns familiar from Earth, where solutions and technologies with the potential to enhance quality of life are available only to the privileged. In this respect, the status of the benefits potentially associated with human space exploration and exploitation resembles that of human enhancement through biomedical means. In the case of human enhancements, which may require expensive technologies, the work on them, as well as their application, will be determined by, and directed to, privileged audiences in terms of gender, race, or class. In the case of space exploration, this is even more noticeable due to the costly and demanding nature of the technical infrastructure. Current space tourism flights illustrate that space is only accessible to influential and wealthy individuals. This practice conflicts with the concepts of space being open to all, accessible, and based on sustainability and equal opportunity (Schwartz 2020). It is hard to believe that with advances in space technology enabling more advanced exploitation of space, whether through space tourism or space mining, space agencies and especially private space companies will invite and help those most in need. These trends are worrying and suggest that the deepening environmental crisis and the superimposed population crisis will increase the exclusion of the poorest and the isolation of those living in richer countries. Some studies show that space exploitation, including space mining, can positively affect sustainability on Earth (Fleming et al. 2023).

As a counterargument, one can mention the alleged specificity of space missions, which can be scientific and technological. This highly technical profile will make the presence of qualified people inevitable, which may shift the balance from the pool of those privileged in terms of social status and wealth to the pool of those privileged in terms of education and the skills desired in space. The same is also true for space tourism, whose participants will be present in space only as tourists and not as explorers, so it may not necessarily affect the composition of the pool of people who will populate space. Space tourists could lead space exploration stakeholders to focus their efforts primarily on tourism, rather than on other types of activities that could potentially be more useful to humanity than space tourism.

The evaluation of space tourism, as well as the involvement of private entities in space exploitation, is nevertheless not clear-cut. A fairly popular belief is that the increasing involvement of private entrepreneurs will lead to lower costs of space mission flights and thus make them more accessible to a wider group of people (Gilley 2020, 4).

It is also worth remembering that space settlement is not proposed as a way to save humanity by sending most of Earth's population, north and south, to live in space, at least not for a very long time. If we could send a thousand people a day into space, which, after all, seems completely unrealistic with today's capabilities, it would take us at least twenty-two thousand years to evacuate Earth. It is worth bearing this in mind, both in a feminist and non-feminist evaluation of the idea of space settlement, when considering its usefulness, especially if the resources allocated for its implementation could be used for the needs of people and the environment on Earth.

In summary, then, the problem is not merely the technological constraint that neither our technological capabilities nor, as determined in part by them, the size, speed, and capacity of spacecraft, as well as the estimated surface area of the target astronomical facility and habitat, will allow for the evacuation from Earth in the coming decades, or perhaps hundreds of years, of a significant number of endangered inhabitants. The problem is the lack of solidarity and the exclusion of those most in need, which means that regardless of tech-nological possibilities, this group will perhaps always be excluded.

Antinatalism as the Only Viable Public Policy of the Future?

Thus we come to a repugnant conclusion. The fate of a significant portion of the poorest people living in the Global South, who are already being hit hardest by the negative effects of climate change, seems a foregone conclusion. Humanity has gone too far past the tipping points of the population-to-resource ratio and environmental destruction to be able to take not only preventive but even adaptive action for all of Earth's inhabitants. If life for a sizable portion of the people of the Global South will either be a life not worth living or not even possible due to climate change, what, if anything, should we do?

The problems discussed apply to future humans who will be born in about a hundred years and beyond. Population ethics considers scenarios in which we determine the impact of currently living generations on generations yet to be born (Parfit 1984). Let us assume, for the sake of argument, one of the worst possible future scenarios, and let us assume that at least a few hundred million of the poorest inhabitants of the most affected parts of Earth will have lives not worth living. These people do not yet exist. Knowing that some future humans will have lives not worth living, is it not better that they never come into the world?

This is the conclusion reached by antinatalist philosophy, which, under what is called "philanthropic antinatalism," concludes that because humans suffer, it is better that we do not reproduce (Benatar 2006). Interestingly, this is a classic variant of antinatalism, applied to humans living under normal conditions. Antinatalists assume that the average human life contains more net suffering than net pleasure. One can easily imagine that if such a conclu-sion is drawn for people living, or about to be born, in normal conditions, it should apply all the more to people who are about to be born in extremely poor conditions. As long as a person has not come into the world, she can not only not experience suffering, but she also cannot be aware of what she has lost as a human being, that is, the lost valuable future typical of human beings. In essence, every reproductive policy and ethic aimed at limiting reproduc-tion is about not bringing future humans into the world who will primarily be exposed to suffering, and who will also, through their own reproduction, bring further future humans into the world under even worse environmental conditions than their own. In this sense, philanthropic antinatalism overlaps with misanthropic antinatalism, which advocates abstaining from reproduction because future humans will cause suffering to others.

The antinatalist position is very controversial, as antinatalists themselves are aware. Perhaps it is an extreme philosophy, perhaps the only rational one (Szocik and Häyry 2024a). But in turning to the antinatalist solution, we allow the risk of abandoning the effort to save humanity and to save, if not the entire population, at least as many of the most endangered as possible.

It is worth asking whether antinatalism should be treated solely as a philosophically interesting concept, or whether it should ever be taken seriously as a potential public policy. This question implies another. Is a future world in which antinatalism is introduced as public policy, or at least is seriously considered by politicians, the preferred future? Depending on how one values the issue of human reproduction, the answer may be either negative or affirmative.

Antinatalism can be the public policy of emergencies when humanity is in a critical situation in which, regardless of the actions taken, a large proportion of the population will have lives not worth living. Despite the special justification in the form of "overpopulation" and climate change for considering antinatalism seriously, the motivation for possibly establishing such a public policy is a concern for future people who will have lives of low quality. The motivation for antinatalism, then, is the specifically understood well-being of future humans, which cannot be guaranteed except by preventing those future humans from coming into existence.

Antinatalism understood in this way is consistent with feminist bioethics. Feminist bioethics is rightly skeptical and critical of the idea of antinatalism, given that in a nonideal world, even formally right ideas will be implemented in oppressive ways against marginalized groups. In contrast, with regard to the antinatalism discussed here, as a global program for future generations, feminist bioethics can support such antinatalism understood as a concern for the protection of all humanity. For in the worst-case scenario, it is the quality of life of all humanity that will be greatly diminished, not just that of those already most excluded.

If a hypothetical future human being is not born as a result of an antinatalist policy, then no harm has been done to them, for one who never existed cannot be disadvantaged by continuing their further state of nonexistence. If she were to come into existence in the absence of an antinatalist policy, her coming into existence would of necessity involve harm being done to her, at least as long as a life lived in overcrowded conditions of poverty, heat, and a lack of drinking water and food is regarded as a life not worth living, or at best as a life barely worth living.

A future scenario in which antinatalism is considered seriously as an optional public policy is perhaps an undesirable scenario, but it is rational and has nothing sinister about it (Szocik and Häyry 2024b). If it were to be judged in moral terms, it is a morally superior action than bringing people into the world knowing that their quality of life will be very low.

But establishing antinatalism as public policy in a future of "overpopulation" and climate change requires at least two conditions. First, we must know that future humans whose coming into the world will be prevented by antinatalist policies will certainly have lives barely worth living or not worth living. The analogy with GGE and euthanasia may be useful here. GGE is often seen in moratoria prohibiting its clinical application as a last-resort option that could theoretically be used only if conventional methods were shown to be ineffective. It is worth noting the rule behind this thinking. Let us assume that a similar rule justifies euthanasia. Antinatalism could only be considered as a final option when it is proven that there are no alternatives. To be so certain, it would have to be shown, among other things, that at no stage in a future person's life will there be any chance of improving the person's well-being. It would also have to be assumed that the objective determinants of well-being would outweigh its subjective determinants.

Meeting the first condition requires meeting the next condition. These are the conditions of fairness, justice, equality, and objectivity. It is possible to imagine a future scenario in which antinatalist policies are not driven by the morally good intentions that guide this discussion but rather by morally bad intentions. Such policies may be directed against particular social groups, selected countries, or even particular nations or ethnic minorities, especially as the total fertility rate is higher in certain countries and certain continents. Antinatalist policies could be perceived as targeting specific groups. This is a danger that would have to be eliminated before antinatalism could be seriously considered. This is also the main reason for feminist criticism of the idea of antinatalism.

Antinatalism is not considered in the context of climate change either as mitigation or adaptation. If, even with the stabilization of greenhouse gas emissions, sea level rises and temperature rises still continue for several hundred years (McGuire 2005, 24–25), reducing the population will not halt effects that were initiated in the past. Antinatalism primarily serves to prevent the probably inevitable suffering of millions and perhaps billions of people in the only way possible under conditions of ecological catastrophe, that is, by not creating them.

Antinatalism as public policy has an indelible racist and discriminatory component. This is so even if the actual intentions of political antinatalists are far from racism and exclusion and are merely the result of purely logical considerations. While the feminist critique of antinatalism as a public policy against the poorest and excluded is valid, the motivation remains an unexplained element. It is not necessarily a deliberately racist motivation; however, this is the case in many situations where feminism shows that even if antinatalism is no longer promoted at the global level as the dominant population policy, many categories of women are targeted as an undesirable category of people (women in prisons, HIV-positive women in Africa and Latin America, or lower-caste women in India) (Bhatia et al. 2020). As the criticism of intersectional feminists points out, the antinatalism presented as a global idea in the past was racist in nature (Sasser 2014).

This discriminatory component of antinatalist public policy stems from its consequences, namely the need to reduce reproduction rates in the Global South, where the total fertility rates are highest and contribute to a systematic decline in quality of life. What is most controversial and problematic is not so much the abstract suspension of reproductive rights and reproductive freedom but the practical temporary prohibition, or at least temporary restriction, of reproduction of populations with the highest total fertility rates.

This equation is inevitable only in the extreme, in the ultimate circumstances. Appropriate social and cultural changes can prevent a course of events in which this equation is inevitable. The alternative to a public policy of antinatalism is to empower women so that they have real reproductive rights and procreative autonomy and can decide for themselves when and how to reproduce. Empowering women is correlated with lowering the total fertility rate (Mulgan 2006, 190–191). However, this is a much more complicated issue because in many cultures women express a desire to have multiple children. High fertility rates are correlated with social and cultural prestige. Feminism in such situations points to the problematic nature of, on the one hand, the risk of oppressing a woman, who for some feminist philosophers may be an unwitting victim of patriarchal domination, and, on the other hand, the need to respect different cultural traditions, especially when they meet with the approval of women in a given community (Okin et al. 1999; Rubio-Marín and Kymlicka 2018). Categories such as false consciousness and adaptive preferences may be useful here to explain those situations in which a woman identifies with the interests of patriarchy or makes the least harmful choice for herself under conditions of limited autonomy.

It is hard to disagree with Tim Mulgan, who argues that the ideal code should take into account reproductive freedom, assume parental obligations and liberal public policy, and not be based on coercion (Mulgan 2006, 195). If we accept a model in which it is the parents and not the state or any third parties who are expected to provide an adequate level of quality of life for future children, the ability to correctly determine the maximum number of offspring who can be provided with a minimum quality of life is crucial here.

If we accept that a huge proportion of the future population will have a life not worth living and at the same time do nothing to prevent this—if any mitigation is realistically possible—then we arrive at the repugnant conclusion that it is better not to be born than to be condemned to a life not worth living. If antinatalism so understood were enacted as a public policy of emergency, it would not only be a rational policy, but also a racist and unjust one regardless of motivations that may be morally good. It would be a policy representing the point of view of the wealthy and privileged, taking away the right to life of future people in the Global South. The statistical fact that the total fertility rate is currently the highest in the Global South does not change the moral assessment of political antinatalism. If we accept the moral validity of international global justice, then the richest countries are responsible for colonizing and exploiting the Global South. To this list must also be added the destruction of the environment, for which those most affected by climate catastrophe today are least to blame.

In sum, injustice and super-exploitation have led to an undesirable future scenario in which perhaps the policy of antinatalism discussed here as a thought experiment may indeed prove to be the optimal solution. But then the people in the Global South most severely affected by climate change should be given an equal right to live in the best-placed-to-survive parts of the world in accordance with the idea of international justice.

Earth and Space Environmental Ethics

Ecofeminism

A branch within feminist ethics with a particular interest in environmental ethics is known as "ecofeminism." Ecofeminism points out that the oppression of women is linked to that of nature, and that feminism should be concerned with environmental issues, while solving environmental problems requires the application of a feminist perspective (Warren 2015).[16] Ecofeminism also emphasizes that the actual source of environmental problems is not anthropocentrism, but androcentrism, that is, a male-centered approach to action

(Tong 2009, 242–243). Chiara Bottici postulates the necessity of authentic freedom for all living beings, not just humans, nor just a certain privileged group. Her "anarchafeminism," which, she stresses, is the same as ecofeminism, grows out of a critique of androcentrism, which today is based on capitalist exploitation and domination (Bottici 2022). This male-centered approach grows out of the masculine primacy of rationality over nature and emotion, out of individualism and an abstract, dichotomous approach to reality (the idea of transcending the natural through rationalization, abstractionism, or finally scientific and technological transformation of the world). This accounts for the oppressive treatment not only of nature but of women as well as animals, who at least since the time of Aristotle have been recognized as beings of lesser moral status than men (Adams and Donovan 1995; Donovan 2012, 206). Ecofeminists assume that a world arranged according to nonmasculine models, taking into account the point of view of groups not allowed to speak, namely women, representatives of other races, and gender nonbinary people, would be a better world for nature and animals (Adams and Gruen 2022).

When discussing bioethical challenges in space, feminism draws attention to why humans intend to explore and exploit space. The bioethical problems awaiting humans in the cosmos will concern human activity in the cosmos, who will transform this cosmos, and probably also destroy it. Therefore, especially for feminism, the discussion of the moral status of individual biomedical challenges in space should include a discussion of the meaning and consequences of just being in space in an environmental context.

The Ideology of Capitalism Once Again: Colonizers' Notion of Wilderness

Chapter 3 discusses "fundamentalist capitalism," one manifestation of which is causing wars (also known as "disaster capitalism"). While chapter 3 describes the negative impact of the ideology of Western expansionism in a social and political context, primarily highlighting the problem of exclusion, here it is worth noting the environmental consequences of this ideology. Two ideas are particularly relevant. One is the concept of property. The other is the notion of wilderness. In practice, these concepts overlap. From the point of view of the European colonizers in America, for example, a territory should not be nobody's or common, but should belong to someone. Feminist critiques of the concepts of property and inheritance, especially within Marxist and socialist feminism, expose the harmfulness of these categories, which can be seen as sources of all social evil, leading to exploitation, abuse, exclusion, violence, and war.

The concept of wilderness was used to justify the conquest and transformation of new territories. The term has a negative connotation and suggests waste, a lack of stewardship, drawing deeper ideological justification from the colonizers' sense of civilizational superiority over the native population. As a consequence, European colonizers conquered and transformed "wild" territories into territories that could be profitable. Ghosh describes this ideology by pointing out, among other things, the differences between European colonial and native Indian understandings of territory and native peoples. For Europeans, the land and its people were savages to be developed. For the indigenous people, these lands were simply tame and full of prosperity (Ghosh 2021, 63–65). American Indians' philosophy of time and perception of reality is circular, not linear, and is based on learning from nature, not conquering it (Wieczorek 2023). The category of wilderness is replete with analogies to racial categories, where black people only came to recognize otherness and its negative social and economic consequences for them after the arrival of white colonizers. In contrast, after the seizure of indigenous territories and the exploitation of people of color, racist institutions and racist mechanisms led to the exposure of mainly people of color to pollution and the negative effects of climate change (Tuana 2023).

Thinking of land in terms of property to be exploited also applies to space. As Linda Billings argues, this understanding of space is particularly characteristic of US politics, replicated by successive presidents such as Ronald Reagan, George W. Bush, Barack Obama, and Donald Trump. The ideological origins of this thinking go back to the Christian Puritanism of the English colonizers of North America in the seventeenth century. As Billings argues, since further US territorial expansion is no longer possible on Earth, space has become the natural new (and only) direction for territorial expansion (Billings 2020, 246–247).[17] This is an inherent feature of American thinking about space, determined by the aforementioned ideology combining nationalism, militarism and colonialism. The essence of this American philosophy of space conquest is perfectly expressed by Rubenstein as follows:

> And there's the US government, making up rules as it goes along and hoping the UN doesn't figure out a way to object. This American astronautic bravado knows no party lines; in fact, the only major Trump-era objectives the Biden administration has retained are (1) the creation of a space force to wage orbital warfare, and (2) the settlement of the moon and Mars to enact what Donald Trump called America's "manifest destiny in the stars." (Rubenstein 2022, ix–x)

It is worth bearing in mind here the different religious traditions and, more broadly, the different cultural outlook on many issues. Both Native Americans and Africans look at nature, property, land, and also the cosmos differently. These perspectives are in contrast to the philosophies of conquest, colonization, and ownership on which European philosophical and religious culture is based (Billings 1997, 2015; Traphagan 2020; Impey 2021).[18]

Environmental problems, like those attributed to "overpopulation," are in practice a consequence of capitalism, whose internal dynamics favor over-exploitation and overconsumption. The consequence of this dynamic is unequal accumulation of wealth. This thus shows that environmental problems are not a natural consequence of increasing numbers of people, but a result of the way wealth is accumulated, which could follow a different, more equality-promoting economy (Fletcher et al. 2014).

Militarization as a Source of Environmental Destruction

The association of militarism, war, and the army with men and masculinity is not just stereotypical. It is a statistically validated correlation. The vast majority of soldiers in modern armies are men; historically, they were almost exclusively men (Benatar 2012). Men's way of solving problems is often based on destructive behavior using violence, weapons, and chemicals (Cribb 2019, 201). If we relate this regularity to the fact that men predominate among criminals who commit the most serious crimes, there are strong reasons to suppose that aggression and violence manifested, among other things, in militarism is culturally and socially, and perhaps also psychologically, related to men.

One should be cautious about drawing such conclusions about supposed differences between female and male approaches to aggression, violence, peace, and nature. The differences that do exist can be, and probably are, due to culture, not biology. Essentialism, or difference feminism, is the exception, not the rule, in feminist philosophy (Hay 2020), although some feminists emphasize the separateness of women from men and attribute to women better—that is, more life-friendly—psychological and moral aptitudes (Andolsen et al. 1987, xvi). It is also pointed out that the conception of women as nonviolent is more a result of their victimization through oppression and violence by men. The link between motherhood and anti-violence is also questioned. Moreover, feminists stress the need for women to be able to respond to male violence with violence (Ruether 1987, 68). But even if the link between aggression and violence and masculinity is purely cultural, this does not change the fact that this correlation occurs along with all of its negative effects such as the aforementioned militarism, domestic violence against women, and warfare (Ruether 1987, 70–71).

Militarization has one particularly negative effect, namely environmental pollution. Militarization is believed to be the largest global source of environmental pollution. The Pentagon is the largest single consumer of energy in the world. The US military consumed more oil during one year of war in Iraq than the annual oil consumption of 180 million people in Bangladesh (Ghosh 2021, 122–123). As Ghosh (2021, 128) adds, the same militarism that destroys the environment has been evident in disaster recovery in recent decades, accustoming the public to progressive militarization.

As long as militarism is associated with what is masculine—even if only because of historical and cultural coincidence—we can expect that the masculine way of solving problems and pursuing policies based on military expansion will lead to further suffering, wars, and environmental destruction. There is no reason to assume that these mechanisms will not apply to expansion in space. Military competition between spacefaring countries may be taking place on Earth long before these countries achieve combat capabilities in space.

Justification for Human Space Missions

Feminism pays special attention to the justification for taking these actions, which carry the risk of discrimination and oppression, as well as destruction. All these risks are present in the case of space exploration. With regard to environmental issues, there are a number of risks associated with human space exploration. Space missions produce environmental pollution both on Earth and in space. The presence of humans on another astronomical object can lead to contamination, which could make it difficult, if not impossible, to determine the genesis of life on a space object, should traces of life be detected but their possible Earth origin not identified. The justification for the presence of humans in space is therefore of vital importance. If human presence could be replaced by robots, there is a strong environmental rationale in favor of robotic missions. As Donald Goldsmith and Martin Rees note, accomplishing scientific goals, for example on Mars, along with the opportunity to enjoy the sights and experience the Martian landscape, could be accomplished by robots, which would be more economical and environmentally friendly than human missions (Goldsmith and Rees 2022, 93).

In conclusion, the feminist bioethics of space exploration will operate in an environment that, at least from today's perspective, may be one based on nationalism, capitalism, militarism, and colonialism on the one hand, and on continuous exploitation leading to environmental degradation on the other. In assessing the ethical status of the various biomedical procedures in space analyzed in the previous chapters, feminism should first and foremost, as a holistic philosophy, expose the mechanisms of destruction, oppression, and exploitation, as well as raising questions about the meaningfulness and legitimacy of such a model of space exploration as is emerging from the current capitalist and militaristic space policy.

Notes

1. Antinatalism is a philosophical form of narrative on the idea of "zero population growth," popularized especially by Paul Ehrlich regarding the "dangers" of unbridled population growth (Ehrlich 1974).
2. For more on population ethics in the context of climate change, see Broome (2022); Conly (2022) and Olsaretti (2022).
3. The Global North's postcolonial approach to the Global South is also evident in the renewable energy transition plans, which guarantee the North's retention of energy privilege, while assuming the use of the South as a source of land for guaranteeing the North's energy privilege (Hickel and Slamersak 2022).
4. Although Khader offers an objective definition of adaptive preferences, it does not completely exclude the constant problem of the risk of applying a paternalistic approach or considering some values more important than those of other cultural circles (Khader 2011). It is also worth bearing in mind in this context the constant ambiguity of human life emphasized by Beauvoir (Kruks 2012).
5. Plato made a similar point in the *Republic*. Some people would have the job of raising the children of women rulers (Plato 1993).
6. A more radical concept is the postulate of nonprocreation, treated as a specific rule of a more general principle, namely the obligation to reduce the individual carbon footprint. Importantly, and in line with the idea of feminist social justice, Trevor Hedberg (2019) addresses this moral obligation to those who live carbon footprint-intensive lifestyles, i.e., residents of developed countries.
7. Within the philosophical and ethical discussion of the idea of space settlement, one of the options under consideration is the concept known as "embryo space settlement," based on ectogenesis (Edwards 2021, 2023). I presented a critique of this concept in Szocik (2021).
8. However, if the preservation of the human species proves to be a space refuge goal, instead of antinatalism, one can use, as Kendal suggests, either global lottery or stratified random sample strategies as the most equitable (2023, 300).
9. Although desertification is progressing in some areas, other areas are seeing an increase in plant presence, particularly in China and India. This is an effect known as the "greening Earth," caused mainly by climate change and CO_2 emissions (Chen et al. 2019).

10. It is estimated that in the twenty-first century alone, rising sea levels could force migration up from seventeen to seventy-two million people. Cost-benefit calculations are being considered to determine which strategy is less costly—protecting the inhabited coastline or evacuating the population (Lincke and Hinkel 2021).

11. It is estimated that as many as 96 percent of victims of natural disasters and environmental change are inhabitants of developing countries (McGuire 2005, 22), which justifies the prediction that in the future it will be the poorest who will suffer even more negative effects of climate change.

12. See the review article by Simon J. Beard and colleagues in which the authors discuss the effects of global warming in terms of global catastrophic risk. Among other things, they cite the views of Karin Kuhlemann, who indicates that climate change will cause a slow decline in well-being. Societies will not be able to guarantee well-being (Beard et al. 2021).

13. Although terms such as "repugnant conclusion," "life barely worth living," and "life not worth living" are inspired by the philosophy of Derek Parfit (1984), in this chapter they are used outside the framework of utilitarian thinking considered by Parfit.

14. As Serene J. Khader argues, Western feminism reinforces the harm done by colonialism. Khader proposes a universalism that will be anti-imperialist and that will take into account the specific cultural context of women outside of Western civilization. Khader argues that the task of feminism is to change power relations, not to postulate a particular set of individual values and goods proposed as objective (universal), which is specific to Western missionary feminism. Feminism can be compatible with various cultural and social forms as long as they serve to combat sexist oppression (Khader 2019, 1–7, 21–25, 143). Western feminism, however, manifests a tendency toward paternalistic and imperialistic treatment of women in non-Western cultures ("colonial feminism"). This leads to the reinforcement of patriarchal and capitalist narratives, as seen in the rhetoric used to justify US and UK military intervention in Iraq and Afghanistan. Consequently, the oppression of women in non-Western cultures is seen as an effect of local cultural conditions rather than one of colonialism (Phipps 2020, 48–49; see also Pitts et al. 2020).

15. There are technological solutions, which, as Gerard O'Neill assumed, could involve space settlements building solar power satellites for which lunar materials could be used. Solar shields could also be created to control the temperature of Earth (Munevar 2023). These solutions require a sufficiently long time to be realized, and they are not solutions for the near future.

16. The above understanding of ecofeminism is sufficient for the present topic of feminist bioethics of space missions, and the different theoretical perspectives within ecofeminism (Merchant 1995) do not affect the shape of the conclusions derived in this book.

17. See also Billings's critique of the US vision of space policy, based on expansionism, imperialism, capitalism, and militarism (Billings 2006a, 2006b, 2018).

18. See the special issue of *Futures* journal on the social, cultural, civilizational, and ethical challenges around the idea of space settlement (*Futures*, volume 110, June 2019).

Conclusions

The feminist approach applied to philosophical and bioethical considerations of future human space missions is to recognize what can go wrong but also creatively propose an alternative vision for future development. But this recognition differs from nonfeminist considerations by adopting specific explanatory categories appropriate to feminism. These categories are categories such as sex, gender, power, oppression, discrimination, and social justice, which were mentioned in all the chapters of this book. Although the dominant theme of this book is feminist bioethics, a good deal of attention has been paid to issues that are nonbioethical but relevant to a holistic view of the place of humans in the cosmos and the effects of space expansion. The book presents a holistic feminist approach to future human space exploration and exploitation.

As we have seen, feminism, applied to the context of future space missions, usually leads to skeptical conclusions. This skepticism applies to safety, autonomy, freedom, justice, and even the justification for the missions themselves. A similar skepticism characterizes considerations of biomedical technologies, such as the biomedical human enhancement, gene editing, and ARTs discussed in this book. In this respect, the feminist bioethics of space missions differs from the nonfeminist bioethics of space exploration. The latter expresses only limited skepticism, which usually relates to the safety of future participants in such expeditions. For feminism, the safety of mission participants is only one of a number of issues under consideration. Because feminist bioethics recognizes and analyzes many more factors, including those related to social and power structures, it remains more skeptical and ambiguous in its ethical assessment of space missions than the usually optimistic nonfeminist bioethics of space exploration.

It is worth bearing in mind, however, that the aforementioned feminist skepticism does not necessarily lead to a negation of either the sense of space missions or that of biomedical procedures in space. This skepticism is a methodological tool that serves to spot potential areas of risk, usually overlooked by nonfeminist approaches. Reproductive issues are a flagship example. For feminist bioethics, the focus is on the woman, her interests, desires, and

Feminist Bioethics in Space. Konrad Szocik, Oxford University Press. © Oxford University Press 2024.
DOI: 10.1093/9780197691076.003.0008

needs, while for nonfeminist bioethics the focus is on the embryo and fetus. The latter perspective will focus on the proper course of pregnancy and saving humanity, while feminism shows us a number of dangers to the woman related to her potential instrumental exploitation and disregard for her will. A similar shift in perspective applies to the interest of feminist bioethics in the interests of other marginalized groups, such as nonwhite, disabled, and sexually nonbinary people.

In conclusion, it is worthwhile taking into account the sensibilities and optics inherent in feminism in philosophical, ethical, and bioethical reflection on the future development of humanity, the impact of technology, and the place of humans in the cosmos. Over the years, we have slowly learned, with great difficulty, to apply feminist thinking to various areas of our life on Earth. The time has finally come to apply the feminist perspective to our thinking about our future in space as well.

References

Adams, C. J., and J. Donovan (Eds.). 1995. *Animals and women: Feminist theoretical explorations.* Duke University Press.

Adams, C. J., and L. Gruen (Eds.). 2022. *Ecofeminism: Feminist intersections with other animals and the earth.* 2nd ed. Bloomsbury Academic.

Alanen, L., and C. Witt (Eds.). 2004. *Feminist reflections on the history of philosophy.* Kluwer Academic Publishers.

Alcoff, L. M. 2006. *Visible identities: Race, gender, and the self.* Oxford University Press.

Ambrogi, I., L. Brito, and R. L. dos Santos. 2023. Epistemic justice and feminist bioethics in global health. *Journal of medical ethics* 49: 345–346.

Anderson, E. 2020. Feminist epistemology and philosophy of science. In E. N. Zalta (Ed.), *The Stanford encyclopedia of philosophy*, Spring 2020 ed. https://plato.stanford.edu/archives/spr2020/entries/feminism-epistemology/.

Anderson, E. 2022. *Black in white space: The enduring impact of color in everyday life.* University of Chicago Press.

Anderson, E., C. Willett, and D. Meyers. 2021. Feminist perspectives on the self. In E. N. Zalta (Ed.), *The Stanford encyclopedia of philosophy*, Fall 2021 ed. https://plato.stanford.edu/archives/fall2021/entries/feminism-self/.

Andolsen, B. H., C. E. Gudorf, and M. D. Pellauer. 1987. Introduction. In B. H. Andolsen, C. E. Gudorf, and M. D. Pellauer (Eds.), *Women's consciousness, women's conscience: A reader in feminist ethics* (pp. xi–xxvi). Harper & Row.

Angouri, J. 2021. Introduction. Language, gender, and sexuality: Sketching out the field. In J. Angouri and J. Baxter (Eds.), *The Routledge handbook of language, gender and sexuality* (pp. 1–21). Routledge.

Antonsen, E., and M. Van Baalen. 2021. *Comparison of health and performance risk for accelerated Mars missions scenarios.* National Aeronautics and Space Administration, Lyndon B. Johnson Space Center, February.

Appel, S. 2015. Post-feminist puritanism: Teaching (and learning from) the Lowell Offering in the 21st century. *Radical teacher* 102: 43–50.

Arditti, R., R. D. Klein, and S. Minden. 1989. Preface to the 1989 edition. In R. Arditti, R. D. Klein, and S. Minden (Eds.), *Test-tube women: What future for motherhood?* (pp. xi–xxiii). Pandora.

Aristotle. 2002. *Nicomachean ethics.* Trans. C. J. Rowe. Oxford University Press.

Arone, A., T. Ivaldi, K. Loganovsky, S. Palermo, E. Parra, W. Flamini, and D. Marazziti. 2021. The burden of space exploration on the mental health of astronauts: A narrative review. *Clinical neuropsychiatry* 18 (5): 237–246.

Arras, J. D. 2009. The way we reason now: Reflective equilibrium in bioethics. In B. Steinbock (Ed.), *The Oxford handbook of bioethics* (pp. 46–71). Oxford University Press.

Arruzza, C., T. Bhattacharya, and N. Fraser. 2019. *Feminism for the 99 percent: A manifesto.* Verso.

Asch, A., and G. Geller. 1996. Feminism, bioethics, and genetics. In S. M. Wolf (Ed.), *Feminism & bioethics: Beyond reproduction* (pp. 318–350). Oxford University Press.

Astronaut/Cosmonaut Statistics. 2024. https://www.worldspaceflight.com/bios/stats1.php. Accessed April 22, 2024.

Bailey, A. 2021. *The weight of whiteness: A feminist engagement with privilege, race, and ignorance*. Rowman & Littlefield.

Balistreri, M., and S. Umbrello. 2022. Space travel does not constitute a condition of moral exceptionality: That which obtains in space obtains also on Earth! *Medicina e morale* 71 (3): 311–321.

Ball, J. R., and C. H. Evans Jr. (Eds.). 2001. *Safe passage: Astronaut care for exploration missions*. National Academy Press.

Ballantyne, A. 2022. Women in research: Historical exclusion, current challenges and future trends. In W. A. Rogers, C. Mills, and J. L. Scully (Eds.), *Routledge handbook of feminist bioethics* (pp. 251–264). Routledge.

Ballou, M. 1995. Naming the issue. In E. J. Rave and C. C. Larsen (Eds.), *Ethical decision making in therapy: Feminist perspectives* (pp. 42–56). Guilford Press.

Balsamo, A. 1996. *Technologies of the gendered body: Reading cyborg women*. Duke University Press.

Bar On, B.-A. 1993. Marginality and epistemic privilege. In L. Alcoff and E. Potter (Eds.), *Feminist epistemologies* (pp. 83–100). Routledge.

Barnes, E. 2016. *The minority body: A theory of disability*. Oxford University Press.

Barrett Meyering, I. 2022. *Feminism and the making of a child rights revolution. 1969–1979*. Melbourne University Press.

Baumgardner, J., and A. Richards. 2020. *Manifesta: Young women, feminism, and the future*. 20th anniversary ed. Picador.

Baylis, F. 2017. Human germline genome editing and broad societal consensus. *Nature human behavior* 1: 0103.

Baylis, F., M. Darnovsky, K. Hasson, and T. M. Krahn. 2020. Human germ line and heritable genome editing: The global policy landscape. *CRISPR journal* 3 (5): 365–377.

Beard, S. J., L. Holt, A. Tzachor, L. Kemp, S. Avin, P. Torres, and H. Belfield. 2021. Assessing climate change's contribution to global catastrophic risk. *Futures* 127: 102673. https://doi.org/10.1016/j.futures.2020.102673.

Beauchamp, T. L., and J. F. Childress (Eds.). 2013. *Principles of biomedical ethics*. 7th ed. Oxford University Press.

Beauvoir, S. de. 2011. *The second sex*. Trans. Constance Borde and Sheila Malovany-Chevallier. Vintage.

Beck, K. 2021. *White feminism: From the suffragettes to influencers and who they leave behind*. Atria Books.

Bee, B. A., and C. M. Park. 2022. Feminist contributions to climate change research, policy and ethics. In W. A. Rogers, C. Mills, and J. L. Scully (Eds.), *Routledge handbook of feminist bioethics* (pp. 547–559). Routledge.

Benatar, D. 2006. *Better never to have been: The harm of coming into existence*. Oxford University Press.

Benatar, D. 2012. *The second sexism: Discrimination against men and boys*. Wiley-Blackwell.

Benenson, J., C. Webb, and R. Wrangham. 2022. Self-protection as an adaptive female strategy. *Behavioral and brain sciences* 45: E128. https://doi.org/10.1017/S0140525X21002417.

Bennhold, K., and M. Pronczuk. 2022. Poland shows the risks for women when abortion is banned. *New York Times*, June 12, 2022. https://www.nytimes.com/2022/06/12/world/europe/poland-abortion-ban.html.

Betson, J. R., and R. R. Secrest. 1964. Prospective women astronauts selection program: Rationale and comments. *American journal of obstetrics and gynecology* 88 (3): 421–423.

Bhandary, A., and A. R. Baehr (Eds.). 2021. *Caring for liberalism: Dependency and liberal political theory*. Routledge.

Bhatia, R., J. S. Sasser, D. Ojeda, A. Hendrixson, S. Nadimpally, and E. E. Foley. 2020. A feminist exploration of "populationism": Engaging contemporary forms of population control. *Gender, place & culture* 27 (3): 333–350.

Bickford, A. 2020. *Chemical heroes: Pharmacological supersoldiers in the US military*. Duke University Press.

Billings, L. (Ed.). 2021. *50 years of solar system exploration: Historical perspectives*. National Aeronautics and Space Administration, Office of Communications, NASA History Division.

Billings, L. 1997. Frontier days in space: Are they over? *Space policy* 13 (3): 187–190.

Billings, L. 2006a. Exploration for the masses? Or joyrides for the ultra-rich? Prospects for space tourism. *Space policy* 22 (3): 162–164.

Billings, L. 2006b. To the moon, Mars, and beyond: Culture, law, and ethics in space-faring societies. *Bulletin of science, technology & society* 26 (5): 430–437.

Billings, L. 2015. Space cowboys. *Scientific American* 313 (2): 12.

Billings, L. 2018. A US Space Force? A very bad idea! *Theology and science* 16 (4): 385–387.

Billings, L. 2020. Earth, life, space: The social construction of the biosphere and the expansion of the concept into outer space. In K. C. Smith and C. Mariscal (Eds.), *Social and conceptual issues in astrobiology* (pp. 239–262). Oxford University Press.

Billings, L. 2023. Neoliberalism: Problematic. Neoliberal space policy? Extremely problematic. In J. S. Johnson-Schwartz, L. Billings, and E. Nesvold (Eds.), *Reclaiming space: Progressive and multicultural visions of space exploration* (pp. 25–36). Oxford University Press.

Birke, L. 1986. *Women, feminism and biology: The feminist challenge*. Wheatsheaf Books.

Bleier, R. (Ed.).1986. *Feminist approaches to science*. Pergamon Press.

Block, J. 2019. *Everything below the waist: Why health care needs a feminist revolution*. St. Martin's Press.

Bonilla-Silva, E. 2022. *Racism without racists: Color-blind racism and the persistence of racial inequality in America*. 6th ed. Rowman & Littlefield.

Bordo, S. 1993. *Unbearable weight: Feminism, Western culture, and the body*. University of California Press.

Bottici, C. 2022. *Anarchafeminism*. Bloomsbury Academic.

Bowden, M. 2021. Deepening futures methods to face the civilisational crisis. *Futures* 132: 102783. https://doi.org/10.1016/j.futures.2021.102783.

Brittan, A., and M. Maynard. 1984. *Sexism, racism, and oppression*. Blackwell.

Broome, J. 2004. *Weighing lives*. Oxford University Press.

Broome, J. 2022. Climate change and population ethics. In G. Arrhenius, K. Bykvist, T. Campbell, and E. Finneron-Burns (Eds.), *The Oxford handbook of population ethics* (pp. 393–406). Oxford University Press.

Brown, F. L., and L. A. Keefer. 2020. Antinatalism from an evolutionary psychological perspective. *Evolutionary psychological science* 6: 283–291.

Brown, P. 2004. *Eve: Sex, childbirth and motherhood through the ages*. Summersdale.

Brundtland, G. H. 1993. Gro Harlem Brundtland on population, environment, and development. *Population and development review* 19 (4): 893–899.

Brundtland, G. H. 1994. Empowering women: The solution to a global crisis. *Environment* 36 (10): 16–20.

Buchanan, A., and R. Powell. 2018. *The evolution of moral progress: A biocultural theory*. Oxford University Press.

Budolfson, M., and D. Spears. 2021. Population ethics and the prospects for fertility policy as climate mitigation policy. *Journal of development studies* 57 (9): 1499–1510.

Burch, S., A. Kafer. 2010. Introduction: Interventions, investments, and intersections. In S. Burch and A. Kafer (eds.), *Deaf and disability studies: Interdisciplinary perspectives* (pp. xiii–xxvii). Gallaudet University Press.

Butler, J. 1990. *Gender trouble: Feminism and the subversion of identity*. Routledge.

Bystranowski, P., V. Dranseika, and T. Żuradzki. 2022. Half a century of bioethics and philosophy of medicine: A topic-modeling study. *Bioethics* 6: 902–925.

Cafaro, P., P. Hansson, and F. Götmark. 2022. Overpopulation is a major cause of biodiversity loss and smaller human populations are necessary to preserve what is left. *Biological conservation* 272: 109646.

Cameron, D. 2007. *The myth of Mars and Venus.* Oxford University Press.

Cameron, D. 2020. *Feminism: A brief introduction to the ideas, debates, and politics of the movement.* University of Chicago Press.

Cameron, J. J. 2022. Radical feminist analysis of heterosexuality. In B. D. Earp, C. Chambers, and L. Watson (Eds.), *The Routledge handbook of philosophy of sex and sexuality* (pp. 180–192). Routledge.

Carse, A. L., and H. Lindemann Nelson. 1999. Rehabilitating care. In A. Donchin and L. M. Purdy (Eds.), *Embodying bioethics: Recent feminist advances* (pp. 17–31). Rowman & Littlefield.

Carter, G. L., and M. L. Fisher. 2017. Conclusion. In M. L. Fisher (Ed.), *The Oxford handbook of women and competition* (pp. 801–810). Oxford University Press.

Cavaliere, G. 2018. Genome editing and assisted reproduction: Curing embryos, society or prospective parents? *Medicine, health care and philosophy* 21: 215–225.

Cavaliere, G. 2020a. Ectogenesis and gender-based oppression: Resisting the ideal of assimilation. *Bioethics* 34: 727–734.

Cavaliere, G. 2020a. Gestation, equality and freedom: Ectogenesis as a political perspective. *Journal of medical ethics* 46: 76–82.

Cavaliere, G. 2020b. The problem with reproductive freedom: Procreation beyond procreators' interests. *Medicine, health care and philosophy* 23: 131–140.

Chaddock, N., and B. Hinderliter. 2020. *Antagonizing white feminism: Intersectionality's critique of women's studies and the academy.* Lexington Books.

Changela, H., E. Chatzitheodoridis, A. Antunes, et al. 2021. Mars: New insights and unresolved questions. *International journal of astrobiology* 20 (6): 394–426.

Chen, C., T. Park, X. Wang, et al. 2019. China and India lead in greening of the world through land-use management. *Nature sustainability* 2: 122–129.

Chon Torres, O. A. 2022. Expansion of humanity in space: Utopia or dystopia? In C. S. Cockell (Ed.), *The institutions of extraterrestrial liberty* (pp. 64–70). Oxford University Press.

Cockell, C. S. (Ed.). 2015a. *Human governance beyond Earth: Implications for freedom.* Springer.

Cockell, C. S. (Ed.). 2015b. *The meaning of liberty beyond Earth.* Springer.

Cockell, C. S. (Ed.). 2016. *Dissent, revolution and liberty beyond Earth.* Springer.

Cockell, C. S. (Ed.). 2022. *The institutions of extraterrestrial liberty.* Oxford University Press.

Code, L. S. 1993. Taking subjectivity into account. In L. Alcoff and E. Potter (Eds.), *Feminist epistemologies* (pp. 15–48). Routledge.

Code, L. S., S. Mullett, and C. Overall. 1988. *Feminist perspectives: Philosophical essays on method and morals.* University of Toronto Press.

Collins, P. H. 2015. Intersectionality's definitional dilemmas. *Annual review of sociology* 41: 1–20.

Conly, S. 2022. Overpopulation and individual responsibility. In G. Arrhenius, K. Bykvist, T. Campbell, and E. Finneron-Burns (Eds.), *The Oxford handbook of population ethics* (pp. 430–443). Oxford University Press.

Cook, R. J. 2002. International human rights and women's reproductive health. In S. Sherwin and B. Parish (Eds.), *Women, medicine, ethics, and the law* (pp. 37–50). Ashgate.

Coole, D. 2021. The toxification of population discourse: A genealogical study. *Journal of development studies* 57 (9): 1454–1469.

Corea, G. 1988. *The mother machine: Reproductive technologies from artificial insemination to artificial wombs.* Women's Press.

Corea, G. 1989. Egg snatchers. In R. Arditti, R. Duelli Klein, and S. Minden (Eds.), *Test-tube women: What future for motherhood?* (pp. 37–51). Pandora.

Crasnow, S. 2020. Feminist perspectives on science. In E. N. Zalta (Ed.), *The Stanford encyclopedia of philosophy*, Winter 2020 ed. https://plato.stanford.edu/archives/win2020/entries/feminist-science/.

Crenshaw, K. 1991. Mapping the margins: Intersectionality, identity politics, and violence against women of color. *Stanford law review* 41: 1241–1298.

Cribb, J. 2019. *Food or war.* Cambridge University Press.

Davenport, C. 2018. *The space barons: Elon Musk, Jeff Bezos, and the quest to colonize the cosmos.* PublicAffairs.

Davis, D.-A. 2019. *Reproductive injustice: Racism, pregnancy, and premature birth.* New York University Press.

Davis, J. K. 2018. *New Methuselahs: The ethics of life extension.* MIT Press.

Dawson, L. 2021. *The politics and perils of space exploration: Who will compete, who will dominate?* 2nd ed. Springer.

De La Torre, G. D., B. van Baarsen, F. Ferlazzo, N. Kanas, K. Weiss, S. Schneider, and I. Whiteley. 2012. Future perspectives on space psychology: Recommendations on psychosocial and neurobehavioural aspects of human spaceflight. *Acta astronautica* 81 (2): 587–599.

de Melo-Martín, I. 2021. The gendered nature of reprogenetic technologies. In S. Crasnow and K. Intemann (Eds.), *The Routledge handbook of feminist philosophy of science* (pp. 289–299). Routledge.

de Melo-Martín, I. 2022a. Genomic technologies: The need for a feminist approach. In W. A. Rogers, C. Mills, and J. L. Scully (Eds.), *Routledge handbook of feminist bioethics* (pp. 276–290). Routledge.

de Melo-Martín, I. 2022b. Reproductive embryo editing: Attending to justice. *Hastings Center report* 52 (4): 26–33.

de Melo-Martín, I. 2023. Feminism and the ethics of reprogenetic technologies. In G. J. Robson and J. Y. Tsou (Eds.), *Technology ethics: A philosophical introduction and readings* (pp. 270–278). Routledge.

DeGrazia, D., and J. Millum. 2021. *A theory of bioethics.* Cambridge University Press.

Deitch, C. 2021. Feminist methodologies. In N. A. Naples (Ed.), *Companion to feminist studies* (pp. 213–230). Wiley-Blackwell.

Delap, L. 2020. *Feminisms: A global history.* University of Chicago Press.

Dembroff, R. 2024. *Real men on top: How patriarchy weaponizes gender.* Oxford University Press.

Diamond, J. 2005. *Collapse: How societies choose to fail or succeed.* Viking Press.

Dickenson, D. 2017. *Property in the body: Feminist perspectives.* Cambridge University Press.

Dodds, S. 2004. Introduction. Integrating global and local perspectives. In R. Tong, A. Donchin, and S. Dodds (Eds.), *Linking visions: Feminist bioethics, human rights, and the developing world* (pp. 1–12). Rowman & Littlefield.

Dodds, S. 2021. Biomedical technologies. In K. Q. Hall and Ásta (Eds.), *The Oxford handbook of feminist philosophy* (pp. 484–494). Oxford University Press.

Donchin, A., and J. L. Scully. 2015. Feminist bioethics. In E. N. Zalta (Ed.), *The Stanford encyclopedia of philosophy*, Winter 2015 ed. https://plato.stanford.edu/archives/win2015/entries/feminist-bioethics/.

Donovan, J. 2012. *Feminist theory: The intellectual traditions.* 4th ed. Continuum.

Dupré, J. 2017. A postgenomic perspective on sex and gender. In D. Livingstone Smith (Ed.), *How biology shapes philosophy: New foundations for naturalism* (pp. 227–246). Cambridge University Press.

Dupré, J., and D. J. Nicholson. 2018. A manifesto for a processual philosophy of biology. In D. J. Nicholson and J. Dupré, *Everything flows: Towards a processual philosophy of biology* (pp. 3–45). Oxford, UK: Oxford University Press.

Dwass, E. 2019. *Diagnosis female: How medical bias endangers women's health.* Rowman & Littlefield.

Edwards, M. R. 2021. Android Noahs and embryo arks: Ectogenesis in global catastrophe survival and space colonization. *International journal of astrobiology* 20 (2): 150–158.

Edwards, M. R. 2023. Blueprint for forever: Securing human far futures with ectogenesis. *Futures* 146: 103085.

Ehrenreich, B., and D. English. 1973. *Witches, midwives & nurses: A history of women healers.* Feminist Press.

Ehrlich, P. R. 1974. Human population and environmental problems. *Environmental conservation* 1 (1): 15–20.

Ehrlich, P. R., and A. H. Ehrlich. 2013. Can a collapse of global civilization be avoided? *Proceedings of the Royal Society B: Biological sciences* 8, 280 (1754): 20122845.

Ellerby, K. 2017. *No shortcut to change: An unlikely path to a more gender-equitable world.* New York University Press.

Elvis, M. 2022. Scarcity in space: Challenges for liberty. In C. S. Cockell (Ed.), *The institutions of extraterrestrial liberty* (pp. 151–172). Oxford University Press.

Eyer, D. 1996. *Motherguilt: How our culture blames mothers for what's wrong with society.* Times Books / Random House.

Eyer, D. E. 1992. *Mother-infant bonding: A scientific fiction.* Yale University Press.

Feder, E. K. 2007. *Family bonds: Genealogies of race and gender.* Oxford University Press.

Federici, S. 2021. *Patriarchy of the wage: Notes on Marx, gender, and feminism.* CA PM Press.

Ferguson, A. 2009. Feminist paradigms of solidarity and justice. *Philosophical topics* 37 (2): 161–177.

Ferrando, F. 2016. Why space migration must be posthuman. In J. Schwartz and T. Milligan (Eds.), *The ethics of space exploration* (pp. 137–152). Springer.

Firestone, S. 1971. *The dialectic of sex: The case for feminist revolution.* Bantam Books.

Fitouchi, L., et al. 2023. Moral disciplining: The cognitive and evolutionary foundations of puritanical morality. *Behavioral and brain sciences* 46: e293. doi:10.1017/S0140525X22002047.

Fleming, M., I. Lange, S. Shojaeinia, and M. Stuermer. 2023. Mining in space could spur sustainable growth. *Proceedings of the National Academy of Sciences* 120 (43): e2221345120.

Fletcher, R., J. Breitling, V. Puleo. 2014. Barbarian Hordes: The Overpopulation Scapegoat in International Development Discourse. *Third world quarterly* 35 (7): 1195–1215.

Flynn, J. 2022. Theory and bioethics. In E. N. Zalta and U. Nodelman (Eds.), *The Stanford encyclopedia of philosophy*, Winter 2022 ed. https://plato.stanford.edu/archives/win2022/entries/theory-bioethics/.

Fourie, C. 2022. "How could anybody think that this is the appropriate way to do bioethics?": Feminist challenges for conceptions of justice in bioethics. In W. A. Rogers, J. L. Scully, S. M. Carter, V. A. Entwistle, and C. Mills (Eds.), *The Routledge handbook of feminist bioethics* (pp. 27–42). Routledge, Taylor & Francis Group.

Frame, D., and M. R. Allen. 2008. Climate change and global risk. In N. Bostrom and M. M. Cirković (Eds.), *Global catastrophic risks* (pp. 265–286). Oxford University Press.

Frazer, E., and N. Lacey. 1993. *The politics of community: A feminist critique of the liberal-communitarian debate.* Harvester Wheatsheaf.

Freeman, J. (Ed.). 1979. *Women, a feminist perspective.* Mayfield.

French, M. 1992. *The war against women.* Summit Books.

Friedan, B. 1963. *The feminine mystique.* Norton.

Friedman, M. 2003. *Autonomy, gender, politics.* Oxford University Press.

Friedman, M., and A. Bolte. 2007. Ethics and feminism. In L. Martín Alcoff and E. Feder Kittay (Eds.), *The Blackwell guide to feminist philosophy* (pp. 81–101). Blackwell.

Frye, M. 1983. *The politics of reality: Essays in feminist theory.* Crossing Press.

Garcia, M. 2021. *We are not born submissive: How patriarchy shapes women's lives.* Princeton University Press.

Gentile, P. 2020. *Queen of the maple leaf: Beauty contests and settler femininity.* UBC Press.

Gentry, C. E., and L. Sjoberg (Eds.). 2015. *Beyond mothers, monsters, whores: Thinking about women's violence in global politics.* Zed Books.

Genz, S., and B. A. Brabon. 2009. *Postfeminism: Cultural texts and theories.* Edinburgh University Press.

Gheaus, A. 2019. More co-parents, fewer children: Multiparenting and sustainable population. *Essays in philosophy* 20 (1): eP1630.

Ghosh, A. 2021. *The nutmeg's curse: Parables for a planet in crisis.* University of Chicago Press.

Gilley, J. 2020. *Space civilization: An inquiry into the social questions for humans living in space.* Lexington Books.

Gilligan, C. 1997. Getting civilized. In A. Oakley and J. Mitchell (Eds.), *Who's afraid of feminism? Seeing through the backlash* (pp. 13–28). Hamish Hamilton.

Giordano, S. 2010. Do we need (bio)ethical principles? In M. Häyry, T. Takala, P. Herissone-Kelly, and G. Árnason (Eds.), *Arguments and analysis in bioethics* (pp. 37–49). Rodopi.

Goldsmith, D., and M. Rees. 2022. *The end of astronauts: Why robots are the future of exploration.* Belknap Press of Harvard University Press.

Goswami, N. 2019. *Subjects that matter: Philosophy, feminism, and postcolonial theory.* State University of New York Press.

Gowaty, P.A. 1992. Evolutionary biology and feminism. *Human nature* 3: 217–249.

Gowdy, J. 2020. Our hunter-gatherer future: Climate change, agriculture and uncivilization. *Futures* 115: 102488, https://doi.org/10.1016/j.futures.2019.102488.

Grasswick, H. 2018. *Feminist social epistemology.* In E. N. Zalta (Ed.), *The Stanford encyclopedia of philosophy*, Fall 2018 ed. https://plato.stanford.edu/archives/fall2018/entries/feminist-social-epistemology/.

Graves, J. L., Jr., and A. H. Goodman. 2022. *Racism, not race: Answers to frequently asked questions.* Columbia University Press.

Gray, P. B., and J. R. Garcia. 2013. *Evolution and human sexual behavior.* Harvard University Press.

Greaves, H. 2019. Climate change and optimum population. *The monist* 102 (1): 42–65.

Green, B. P. 2021. *Space ethics.* Rowman & Littlefield.

Gres, T., E. Richardson, M. Choudhary, H. Haile, and H. A. P. Cano. 2022. Astronauts with disabilities: A dream becoming reality for a bigger part of humanity. 73rd International Astronautical Congress, Paris, France, 18–22 September 2022.

Grigg, A. J., and A. Kirkland. 2015. Health. In L. Disch and M. Hawkesworth (Eds.), *The Oxford handbook of feminist theory* (pp. 326–345). Oxford University Press.

Grimshaw, J. 1986. *Philosophy and feminist thinking.* University of Minnesota Press.

Griscom, J. L. 1987. On healing the nature/history split in feminist thought. In B. H. Andolsen, C. E. Gudorf, and M. D. Pellauer (Eds.), *Women's consciousness, women's conscience: A reader in feminist ethics* (pp. 85–98). Harper & Row.

Gupta, K. 2020. *Medical entanglements: Rethinking feminist debates about healthcare.* Rutgers University Press.

Hall, K. Q. (Ed.). 2011. *Feminist disability studies.* Indiana University Press.

Hall, M. C. 2013. Reconciling the disability critique and reproductive liberty: The case of negative genetic selection. *IJFAB: International journal of feminist approaches to bioethics* 6 (1): 121–143.

Hanson, R. 2008. Catastrophe, social collapse, and human extinction. In N. Bostrom and M. M. Ćirković (Eds.), *Global catastrophic risks* (pp. 363–377). Oxford University Press.

Haraway, D. J. 2015. Anthropocene, Capitalocene, Plantationocene, Chthulucene: Making kin. *Environmental humanities* 6 (1): 159–165.

Haraway, D. J. 1991. A cyborg manifesto: Science, technology, and socialist-feminism in the late twentieth century. In D. J. Haraway (Ed.), *Simians, cyborgs, and women: The reinvention of nature* (pp. 149–181). Routledge.

Hardin, G. 1968. The tragedy of the commons. *Science* 162 (3859): 1243–1248.

Harding, S., and M. B. Hintikka (Eds.). 2003. *Discovering reality: Feminist perspectives on epistemology, metaphysics, methodology, and philosophy of science.* Springer.

Harrison, L. 2016. *Brown bodies, white babies: The politics of cross-racial surrogacy.* New York University Press.

Haslanger, S. 2020. Why I don't believe in patriarchy: Comments on Kate Manne's down girl. *Philosophy and phenomenological research* 101 (1): 220–229.

Hay, C. 2020. *Think like a feminist: The philosophy behind the revolution.* Norton.

Häyry, M. 2010. An analysis of some arguments for and against human reproduction. In M. Häyry, T. Takala, P. Herissone-Kelly, and G. Árnason (Eds.), *Arguments and analysis in bioethics* (pp. 167–175). Rodopi.

Hedberg, T. 2019. The duty to reduce greenhouse gas emissions and the limits of permissible procreation. *Essays in Philosophy* 20 (1): eP1628.

Heinicke, C., M. Kaczmarzyk, B. Tannert, et al. 2021. Disability in space: Aim high. *Science* 372: 1271–1272.

Hendl, T., and T. K. Browne. 2022. Gender. Ongoing debates and future directions. In W. A. Rogers, J. L. Scully, S. M. Carter, V. A. Entwistle, and C. Mills (Eds.), *The Routledge handbook of feminist bioethics* (pp. 151–166). Routledge, Taylor & Francis Group.

Hendrixson, A., D. Ojeda, J. S. Sasser, S. Nadimpally, E. E. Foley, and R. Bhatia. 2020. Confronting populationism: Feminist challenges to population control in an era of climate change. *Gender, place & culture* 27 (3): 307–315.

Hester, H. 2018. *Xenofeminism.* Polity Press.

Heywood, L., and J. Drake (Eds.). 1997. *Third wave agenda: Being feminist, doing feminism.* University of Minnesota Press.

Heywood, L., and J. Drake. 2004. "It's all about the Benjamins": Economic determinants of the third wave feminism in the United States. In S. Gillis, G. Howie, and R. Munford (Eds.), *Third wave feminism: A critical exploration* (pp. 13–23). Palgrave Macmillan.

Hickel, J., and A. Slamersak. 2022. Existing climate mitigation scenarios perpetuate colonial inequalities. *Lancet planet health* 6 (7): e628–e631.

Hill, M., K. Glaser, and J. Harden. 1995. A feminist model for ethical decision making. In E. J. Rave and C. C. Larsen (Eds.), *Ethical decision making in therapy: Feminist perspectives* (pp. 18–37). Guilford Press.

Ho, A. 2014. Choosing death: Autonomy and ableism. In A. Veltman and M. Piper (Eds.). *Autonomy, oppression, and gender* (pp. 326–349). Oxford University Press.

Hoagland, S. L. 1990. Lesbian ethics and female agency. In J. Allen (Ed.), *Lesbian philosophies and cultures* (pp. 275–291). State University of New York Press.

Holland, N. J. 1990. *Is women's philosophy possible?* Rowman & Littlefield.

Holmes, H. Bequaert. 1999. Closing the gaps: An imperative for feminist bioethics. In A. Donchin, and L. M. Purdy (Eds.), *Embodying bioethics: Recent feminist advances* (pp. 45–63). Rowman & Littlefield.

hooks, b. 2001. Black women: Shaping feminist theory. In Kum-Kum Bhavnani (Ed.), *Feminism and "race"* (pp. 33–39). Oxford University Press.

Hrdy, S. B. 1999. *The woman that never evolved.* Harvard University Press.

Hubbard, R. 1989. Personal courage is not enough: Some hazards of childbearing in the 1980s. In R. Arditti, R. Duelli Klein, and S. Minden (Eds.), *Test-tube women: What future for motherhood?* (pp. 331–355). Pandora.

Hunt, G. 2017. Intersectionality: Locating and critiquing internal structures of oppression within feminism. In C. Hay (Ed.), *Philosophy: Feminism* (pp. 121–138). Macmillan Reference USA.

Hutchison, K. 2022. Feminist epistemology. In W. A. Rogers, J. L. Scully, S. M. Carter, V. A. Entwistle, and C. Mills (Eds.), *The Routledge handbook of feminist bioethics* (pp. 43–57). Routledge, Taylor & Francis Group.

Impey, C. D. 2021. Science and faith off-Earth. In M. B. Rappaport and K. Szocik (Eds), *The human factor in the settlement of the moon* (pp. 245–255). Springer.

Ingarden, R. 1983. *Man and value.* Trans. Arthur Szylewicz. Catholic University of America Press; Philosophia Verlag.

Jaggar, A. 2000. Feminism in ethics: Moral justification. In M. Fricker and J. Hornsby (Eds.), *The Cambridge companion to feminism in philosophy* (pp. 225–244). Cambridge University Press.

Jenkins, P. 2021. *Climate, catastrophe, and faith: How changes in climate drive religious upheaval.* Oxford University Press.

Jenkins, S. C. 2016. Defining morally considerable life: Toward a feminist disability ethics. In H. Sharp and C. Taylor (Eds.). *Feminist philosophies of life* (pp. 199–216). McGill-Queen's University Press.

Johnson, A. G. 2014. *The gender knot: Unraveling our patriarchal legacy.* Temple University Press.

Jones, C. 2021. Getting past Cassandra: 21 C Slaughter. *Futures* 132: 102790.

Juengst, E., and D. Moseley. 2019. *Human enhancement.* In E. N. Zalta (Ed.), *The Stanford encyclopedia of philosophy*, Summer 2019 ed. https://plato.stanford.edu/archives/sum2019/entries/enhancement/.

Kanas, N. 2011. From Earth's orbit to the outer planets and beyond: Psychological issues in space. *Acta astronautica* 68 (5–6): 576–581.

Kanas, N. 2014. Psychosocial issues during an expedition to Mars. *Acta astronautica* 103: 73–80.

Kendal, E. 2018. Utopian literature and bioethics: Exploring reproductive difference and gender equality. *Literature and medicine* 36 (1): 56–84.

Kendal, E. 2020. Biological modification as prophylaxis: How extreme environments challenge the treatment/enhancement divide. In K. Szocik (Ed.), *Human enhancements for space missions: Lunar, Martian, and future missions to the outer planets* (pp. 35–46). Springer.

Kendal, E. 2021. Environmental and occupational ethics in early lunar populations: Establishing guidelines for future off-world settlements. In M. B. Rappaport and K. Szocik (Eds.), *The human factor in the settlement of the moon* (pp. 233–243). Springer.

Kendal, E. 2022. Ectogenesis and the ethics of new reproductive technologies for space exploration. In E. Tumilty and M. Battle-Fisher (Eds.), *Transhumanism: Entering an era of bodyhacking and radical human modification* (pp. 211–226). Springer.

Kendal, E. 2023. Desire, duty, and discrimination: Is there an ethical way to select humans for Noah's Ark? In J. S. J. Schwartz, L. Billings, and E. Nesvold (Eds.), *Reclaiming space: Progressive and multicultural visions of space exploration* (pp. 289–302). Oxford University Press.

Kendall, M. 2020. *Hood feminism: Notes from the women that a movement forgot.* Viking.

Khader, S. J. 2011. *Adaptive preferences and women's empowerment.* Oxford University Press.

Khader, S. J. 2019. *Decolonizing universalism: A transnational feminist ethic.* Oxford University Press.

Klein, N. 2007. *The shock doctrine: The rise of disaster capitalism.* Metropolitan Books / Henry Holt.

Kocum, L., D. S. Courvoisier, and S. Vernon. 2017. The buzz on the queen bee and other characterizations of women's intrasexual competition at work. In M. L. Fisher (Ed.), *The Oxford handbook of women and competition* (pp. 719–738). Oxford University Press.

Kolbert, K., and J. F. Kay. 2021. *Controlling women: What we must do now to save reproductive freedom.* Hachette Books.

Kramer, L., and A. Beutel. 2015. *The sociology of gender: A brief introduction.* Oxford University Press.

Kruks, S. 2012. *Simone de Beauvoir and the politics of ambiguity.* Oxford University Press.

Kuhn, T. S. 1962. *The structure of scientific revolutions.* University of Chicago Press.

Kukla, Q. R. 2020. Misogyny and ideological logic. *Philosophy and phenomenological research* 101: 230–235.

Kukla, R., and K. Wayne. 2018. Pregnancy, birth, and medicine. In E. N. Zalta (Ed.), *The Stanford encyclopedia of philosophy*, Spring 2018 ed. https://plato.stanford.edu/archives/spr2018/entries/ethics-pregnancy.

Labude, M. K., V. Xafis, P. S. Lai, and C. Mills. 2022. Vulnerability and the ethics of human germline genome editing. *CRISPR journal* 5 (3): 358–363.

Laing, O. 2021. *Everybody: A book about freedom*. Norton.

Landis, G. A. 2000. An all-woman crew to Mars: A radical proposal. *Space Policy* 16 (3): 167–169.

Le Guin, U. K. 1969. *The left hand of darkness*. Ace Books.

Le Roy, B., C. Martin-Krumm, N. Pinol, F. Dutheil, and M. Trousselard. 2023. Human challenges to adaptation to extreme professional environments: A systematic review. *Neuroscience & biobehavioral reviews* 146: 105054.

LeMoncheck, L. 2002. Philosophy, gender politics, and *in vitro* fertilization: A feminist ethic of reproductive healthcare. In S. Sherwin and B. Parish (Eds.), *Women, medicine, ethics, and the law* (pp. 149–165). Ashgate.

Lewis, S. 2019. *Full surrogacy now: Feminism against family*. Verso.

Liburd, R., and E. Rothblum. 1995. The medical model. In E. J. Rave and C. C. Larsen (Eds.), *Ethical decision making in therapy: Feminist perspectives* (pp. 177–201). Guilford Press.

Lieberman, B., and E. Gordon. 2018. *Climate change in human history: Prehistory to the present*. Bloomsbury Academic.

Lincke, D., and J. Hinkel, J. 2021. Coastal migration due to 21st century sea-level rise. *Earth's future* 9: e2020EF001965. https://doi.org/10.1029/2020EF001965.

Lindemann, H. 2007. Feminist bioethics: Where we've been, where we're going. In L. Martín Alcoff and E. Feder Kittay (Eds.), *The Blackwell guide to feminist philosophy* (pp. 116–130). Blackwell.

Lindemann, H. 2019. *An invitation to feminist ethics*. 2nd ed. Oxford University Press.

Lindemann, H. 2019. Surgeon general's warning: Gender is bad for your health. *Hastings Center report* 49 (6): 3–3.

Lindemann, H. 2021. Feminist bioethical theory: Feminist bioethics. *Routledge encyclopedia of philosophy*. Taylor and Francis. https://www.rep.routledge.com/articles/thematic/feminist-bioethics/v-1.

Lorber, J. 1997. *Gender and the social construction of illness*. Sage Publications.

Lugones, M. 2003. *Pilgrimages = peregrinajes: Theorizing coalition against multiple oppressions*. Rowman & Littlefield.

MacDougall, D. R. 2022. *Righting health policy: Bioethics, political philosophy, and the normative justification of health law and policy*. Lexington Books.

MacKay, K. 2020. The "tyranny of reproduction": Could ectogenesis further women's liberation? *Bioethics* 34: 346–353.

Mackenzie, C. 2014. Three dimensions of autonomy: A relational analysis. In A. Veltman and M. Piper (Eds.), *Autonomy, oppression, and gender* (pp. 15–41). Oxford University Press.

Mackenzie, C. 2021. Relational autonomy. In K. Q. Hall and Ásta (Eds.), *The Oxford handbook of feminist philosophy* (pp. 374–384). Oxford University Press.

Macpherson, C. C. 2016. Why bioethics should address climate change and how it might do so. In C. C. Macpherson (Ed.), *Bioethical insights into values and policy: Climate change and health* (pp. 199–216). Springer.

Maguire, E. 2019. *This is what a feminist looks like: The rise and rise of Australian feminism*. NLA Publishing.

Malka, Gelfand S. 2007. *Daring to care: American nursing and second-wave feminism*. University of Illinois Press.

Mann, B. 2014. *Sovereign masculinity: Gender lessons from the war on terror*. Oxford University Press.

Manne, K. 2018. *Down girl: The logic of misogyny*. Oxford University Press.

Marge, M. 2022. Preparing individuals with disabilities for space travel and habitation. *Disability and health journal* 15 (2): 101228.

Marway, H., and H. Widdows. 2015. Philosophical feminist bioethics: Past, present, and future. *Cambridge quarterly of healthcare ethics* 2 (24): 165–174.

Mason, C. E. 2021. *The next 500 years: Engineering life to reach new worlds*. MIT Press.

Mason, E. 2022. *Feminist philosophy: An introduction*. Routledge.

McFall, John. N.d. European Space Agency. https://www.esa.int/Science_Exploration/ Human_and_Robotic_Exploration/Astronauts/John_McFall. Accessed March 17, 2024.

McGregor, A. J. 2020. *Sex matters: How male-centric medicine endangers women's health and what we can do about it*. Hachette Go.

McGuire, B. 2005. *Global catastrophes: A very short introduction*. Oxford University Press.

McLaren, M. A. 2019. *Women's activism, feminism, and social justice*. Oxford University Press.

McLeod, C. 2022. The right to reproduce. In W. A. Rogers, C. Mills, and J. L. Scully (Eds.), *Routledge handbook of feminist bioethics* (pp. 451–462). Routledge.

McRuer, R. 2006. *Crip theory: Cultural signs of queerness and disability*. New York University Press.

Merchant, C. 1995. *Earthcare: Women and the environment*. Routledge.

Meyers, Diana T. 2002. *Gender in the mirror: Cultural imagery and women's agency*. Oxford University Press.

Mikkola, M. 2016. *The wrong of injustice: Dehumanization and its role in feminist philosophy*. Oxford University Press.

Mikkola, M. 2017. Gender essentialism and anti-essentialism. In A. Garry, S. J. Khader, and A. Stone (Eds.), *The Routledge companion to feminist philosophy* (pp. 168–179). Routledge, Taylor & Francis Group.

Millett, K. 1970. *Sexual politics*. Doubleday.

Milligan, T. 2015. Rawlsian deliberation about space settlement. In C. Cockell (Ed.), *Human governance beyond Earth* (pp. 9–22). Springer.

Milligan, T. 2016. Space ethics without foundations. In J. Schwartz and T. Milligan (Eds.), *The ethics of space exploration* (pp. 125–134). Springer.

Milligan, T. 2022. Inclusion and environmental protection in space. *Brown journal of world affairs* 29 (1): 117–128.

Milligan, T., and J. S. Johnson-Schwartz. 2023. Space ethics. In J. F. Salazar and A. Gorman (Eds.), *The Routledge handbook of social studies of outer space* (pp. 108–120). Routledge.

Mills, C. W. 1997. *The racial contract*. Cornell University Press.

Mills, C. W. 2017. *Black rights / white wrongs: The critique of racial liberalism*. Oxford University Press.

Mingus, M. 2017. Moving toward the ugly: A politic beyond desirability. In L. J. Davis (ed.), *Beginning with disability: A primer* (pp. 137–141). Routledge.

Mohanty, C. T., A. Russo, and L. Torres (Eds.). 1991. *Third world women and the politics of feminism*. Indiana University Press.

Mosko, M. 2018. Feminist ethics. In R. W. Kolb (Ed.), *The Sage encyclopedia of business ethics and society*, 2nd ed., vol. 3 (pp. 1366–1370). Sage Publications.

Mosquera, J. 2022. Disability and population ethics. In G. Arrhenius, K. Bykvist, T. Campbell, and E. Finneron-Burns (Eds.), *The Oxford handbook of population ethics* (pp. 588–614). Oxford University Press.

Mulgan, T. 2006. *Future people: A modern consequentialist account of our obligations to future generations*. Clarendon Press; Oxford University Press.

Munevar, G. 2023. *The dimming of starlight: The philosophy of space exploration*. Oxford University Press.

Murphy, J. 1989. Egg farming and women's future. In R. Arditti, R. Duelli Klein, and S. Minden (Eds.), *Test-tube women: What future for motherhood?* (pp. 68–75). Pandora.

Nelson, L. H. 2017. *Biology and feminism: A philosophical introduction*. Cambridge University Press.

Nesvold, E. 2023. *Off-Earth: Ethical questions and quandaries for living in outer space*. MIT Press.

Newman, L. M. 1999. *White women's rights: The racial origins of feminism in the United States*. Oxford University Press.

Nisha Z. 2021 Technicization of "birth" and "mothering": Bioethical debates from feminist perspectives. *Asian bioethics review* 13 (2): 133–148.

Nussbaum, M. C. 2000. *Women and human development: The capabilities approach*. Cambridge University Press.

Nussbaum, M. C. 2011. *Creating capabilities: The human development approach*. Belknap Press of Harvard University Press.

Nussbaum, M. C., M. Howard, J. Cohen, and S. M. Okin (Eds.). 1999. *Is multiculturalism bad for women?* Princeton University Press.

O'Sullivan, J. N. 2023. Demographic delusions: World population growth is exceeding most projections and jeopardising scenarios for sustainable futures. *World* 4 (3): 545–568.

Oakley, A. 1997. A brief history of gender. In A. Oakley and J. Mitchell (Eds.), *Who's afraid of feminism? Seeing through the backlash* (pp. 29–55). Hamish Hamilton.

Ojeda, D., J. S. Sasser, E. Lunstrum. 2020. Malthus's specter and the Anthropocene. *Gender, place & culture* 27:3: 316–332.

Okin, S. M. 1989. *Justice, gender, and the family*. Basic Books.

Okin, S. M., J. Cohen, M. Howard, M. C. Nussbaum. 1999. *Is multiculturalism bad for women?* Princeton University Press.

Olsaretti, S. 2022. Egalitarian justice, population size, and parents' responsibility for the costs of children. In G. Arrhenius, K. Bykvist, T. Campbell, and E. Finneron-Burns (Eds.), *The Oxford handbook of population ethics* (pp. 407–429). Oxford University Press.

Olufemi, L. 2020. *Feminism, interrupted: Disrupting power. Outspoken*. Pluto Press.

Ortega, M. 2006. Being lovingly, knowingly ignorant: White feminism and women of color. *Hypatia* 21 (3): 56–74.

Ortner, S. B. 2022. Patriarchy. *Feminist anthropology* 3: 307–314.

Paikowsky, D. 2017. *The power of the space club*. Cambridge University Press.

Parfit, D. 1984. *Reasons and persons*. Clarendon Press.

Parr, A. 2021. The gender-climate-injustice nexus. In K. Q. Hall and Ásta (Eds.), *The Oxford handbook of feminist philosophy* (pp. 474–483). Oxford University Press.

Parsons, P. 2020. *Space travel: Ten short lessons*. Johns Hopkins University Press.

Pateman, C. 1988. *The sexual contract*. Polity.

Paton, A. 2022. The surveillance of pregnant bodies in the age of digital health: Ethical dilemmas. In W. A. Rogers, C. Mills, and J. L. Scully (Eds.), *Routledge handbook of feminist bioethics* (pp. 476–485). Routledge.

Phipps, A. 2020. *Me, not you: The trouble with mainstream feminism*. Manchester University Press.

Pitts, A. J., M. Ortega, and J. Medina (Eds.). 2020. *Theories of the flesh: Latinx and Latin American feminisms, transformation, and resistance*. Oxford University Press.

Plank, L. 2019. *For the love of men: A new vision for mindful masculinity*. St. Martin's Press.

Plato. 1993. *Republic*. Trans. R. Waterfield. Oxford University Press.

Prysyazhnyuk, A., and C. McGregor. 2022. Space as an extreme environment—galactic adventures: Exploring the limits of human mind and body, one planet at a time. In T. Cibis and C. McGregor (Eds.), *Engineering and medicine in extreme environments* (pp. 121–141). Springer.

Purdy, L. M. 1996. *Reproducing persons: Issues in feminist bioethics.* Cornell University Press.

Ralston, M. 2021. *Slut-shaming, whorephobia, and the unfinished sexual revolution.* McGill-Queen's University Press.

Räsänen, J., M. Häyry. 2023. Antinatalism—solving everything everywhere all at once? *Bioethics* 37: 829–830.

Rawlinson, M. C. 2001. The concept of a feminist bioethics. *Journal of medicine and philosophy* 26 (4): 405–416.

Rawlinson, M. C. 2016. *Just life: Bioethics and the future of sexual difference.* Columbia University Press.

Rawls, J. 1999. *A theory of justice.* Rev. ed. Harvard University Press.

Ray, K., J. F. Cooper. 2024. The bioethics of environmental injustice: Ethical, legal, and clinical implications of unhealthy environments. *American journal of bioethics* 24 (3): 9–17. https://doi.org/10.1080/15265161.2023.2201192.

Resnik, D. B. 2022. Environmental justice and climate change policies. *Bioethics* 36: 735–741.

Roberts, D. E. 2011. *Fatal invention: How science, politics, and big business re-create race in the twenty-first century.* New Press.

Rogers, W. A., J. L. Scully, S. M. Carter, V. A. Entwistle, and C. Mills. 2022. Introduction. In W. A. Rogers, J. L. Scully, S. M. Carter, V. A. Entwistle, and C. Mills (Eds.), *The Routledge handbook of feminist bioethics* (pp. 1–11). Routledge, Taylor & Francis Group.

Rosser, S. V. (Ed.). 1988. *Feminism within the science and health care professions: Overcoming resistance.* Pergamon Press.

Rosser, S. V. 1994. *Women's health-missing from U.S. medicine.* Indiana University Press.

Rothman, B. K., and M. B. Caschetta. 1995. Treating health: Women and medicine. In J. Freeman (Ed.), *Women: A feminist perspective,* 5th ed. (pp. 65–78). Mayfield.

Rowland, R. 1989. Reproductive technologies: The final solution to the woman question? In R. Arditti, R. Duelli Klein, and S. Minden (Eds.), *Test-tube women: What future for motherhood?* (pp. 356–369). Pandora.

Rubenstein, M. 2022. *Astrotopia: The dangerous religion of the corporate space race.* University of Chicago Press.

Rubio-Marín, R., W. Kymlicka. 2018. *Gender parity and multicultural feminism: Towards a new synthesis.* Oxford University Press.

Rueda, J. 2022. Genetic enhancement, human extinction, and the best interests of posthumanity. *Bioethics* 1–10. https://doi.org/10.1111/bioe.13085.

Ruether, R. R. 1987. Feminism and peace. In B. H. Andolsen, C. E. Gudorf, and M. D. Pellauer (Eds.), *Women's consciousness, women's conscience: A reader in feminist ethics* (pp. 63–74). Harper & Row.

Russell, C. 2022. What makes an antiracist feminist bioethics? In W. A. Rogers, J. L. Scully, S. M. Carter, V. A. Entwistle, and C. Mills (Eds.), *The Routledge handbook of feminist bioethics* (pp. 195–207). Routledge, Taylor & Francis Group.

Samerksi, S. 2015. Pregnancy, personhood, and the making of the fetus. In L. Disch and M. Hawkesworth (Eds.), *The Oxford handbook of feminist theory* (pp. 699–720). Oxford University Press.

Sasser, J. 2014. From darkness into light: Race, population, and environmental advocacy. *Antipode* 46: 1240–1257.

Sasser, J. S. 2018. *On infertile ground: Population control and women's rights in the era of climate change.* New York University Press.

Sasser, J. S. 2023. At the intersection of climate justice and reproductive justice. *WIREs climate change* 15 (4): e860. https://doi.org/10.1002/wcc.860.

Savigny, H. 2020. *Cultural sexism: The politics of feminist rage in the #MeToo era.* Bristol University Press.

Sawin, C. 2021. Bioethics in space exploration. In L. R. Young and J. P. Sutton (Eds.), *Handbook of bioastronautics* (pp. 565–572). Springer.

Saxton, M. 2016. Access to care: The heart of equity in health care. In S. E. Miles-Cohen and C. Signore (Eds.), *Eliminating inequities for women with disabilities: An agenda for health and wellness* (pp. 39–59). American Psychological Association.

Schrogl, K.-U. 2023. Space sustainability and the global commons. *Room: Space journal of Asgardia* 33: 26–29.

Schuller, K. 2021. *The trouble with white women: A counterhistory of feminism.* Bold Type Books.

Schultz, S. 2010. Redefining and medicalizing: NGOs and their innovative contributions to the post-Cairo agenda. In M. Rao and S. Sexton (Eds.), *Markets and Malthus: Population, gender, and health in neo-liberal times* (pp. 173–215). Sage Publications.

Schulz, A. J., and L. Mullings (Eds.). 2006. *Gender, race, class, and health: Intersectional approaches.* Jossey-Bass.

Schwartz, J. S. J. 2020. The accessible universe: On the *choice* to require bodily modification for space exploration. In K. Szocik (Ed.), *Human enhancements for space missions: Lunar, Martian, and future missions to the outer planets* (pp. 201–215). Springer.

Schwartz, J. S. J. 2022. Justice in space: Demanding political philosophy for demanding environments. In C. S. Cockell (Ed.), *The institutions of extraterrestrial liberty* (pp. 411–422). Oxford University Press.

Schwartz, J. S. J., L. Billings, and E. Nesvold (Eds.), 2023. *Reclaiming space: Progressive and multicultural visions of space exploration.* Oxford University Press.

Schwartzman, L. H. 2006. *Challenging liberalism: Feminism as political critique.* Pennsylvania State University Press.

Scott, J. W. 2018. *Sex and secularism.* Princeton University Press.

Scott, R. 2022. Women, assisted reproduction and the "natural." In W. A. Rogers, J. L. Scully, S. M. Carter, V. A. Entwistle, and C. Mills (Eds.), *The Routledge handbook of feminist bioethics* (pp. 463–475). Routledge, Taylor & Francis Group.

Scully, J. L. 2008. *Disability bioethics: Moral bodies, moral difference.* Rowman & Littlefield.

Scully, J. L. 2021. Feminist bioethics. In K. Q. Hall and Ásta (Eds.), *The Oxford handbook of feminist philosophy* (pp. 272–286). Oxford University Press.

Scully, J. L. 2022. Feminist bioethics and disability. In W. A. Rogers, C. Mills, and J. L. Scully (Eds.), *Routledge handbook of feminist bioethics* (pp. 181–194). Routledge.

Sen, G. 2014. *The remaking of social contracts: Feminists in a fierce new world.* Zed Books.

Seymour, N. 2013. Down with people: Queer tendencies and troubling racial politics in antinatalist discourse. In G. Gaard, S. C. Estok, and S. Oppermann (Eds.), *International perspectives in feminist ecocriticism* (pp. 203–218). Routledge.

Shahvisi, A. 2022. Toward an anticolonial feminist bioethics. In W. A. Rogers, J. L. Scully, S. M. Carter, V. A. Entwistle, and C. Mills (Eds.), *The Routledge handbook of feminist bioethics* (pp. 208–221). Routledge, Taylor & Francis Group.

Shanthi, K. 2004. Feminist bioethics and reproductive rights in India. In R. Tong, A. Donchin, and S. Dodds (Eds.), *Linking visions: Feminist bioethics, human rights, and the developing world* (pp. 119–132). Rowman & Littlefield.

Shayler, D. J., and I. A. Moule. 2005. *Women in space: Following Valentina.* Praxis.

Sheffield, C. J. 1995. Sexual terrorism. In J. Freeman (Ed.), *Women: A feminist perspective*, 5th ed. (pp. 1–21.). Mayfield.

Shelhamer, M. J., and G. B. I. Scott. 2021. Future human exploration challenges: An overview. In L. R. Young and J. P. Sutton (Eds.), *Handbook of bioastronautics* (pp. 799–806). Springer.

Sherwin, S. 1992. *No longer patient: Feminist ethics and health care.* Temple University Press.

Sherwin, S. 1996. Feminism and bioethics. In S. M. Wolf (Ed.), *Feminism & bioethics: Beyond reproduction* (pp. 47–66). Oxford University Press.

Sherwin, S. 2001. Moral perception and global visions. *Bioethics* 15: 175–188.

Sherwin, S., and K. Stockdale. 2017. Whither bioethics now? The promise of relational theory. *IJFAB: International journal of feminist approaches to bioethics* 10 (1): 7–29.

Shildrick, M. 1997. *Leaky bodies and boundaries: Feminism, postmodernism and (bio)ethics.* Routledge.

Silvermint, D. 2017. Feminist perspectives on sexism and oppression. In C. Hay (Ed.), *Philosophy: Feminism* (pp. 37–69). Macmillan Reference USA.

Simonstein, F. 2019. Gene editing, enhancing and women's role. *Science and engineering ethics* 25: 1007–1016.

Smith, K. C., and C. Hylkema. 2020. Who's afraid of little green men? Genetic enhancement for off-world settlements. In K. Szocik (Ed.), *Human enhancements for space missions: Lunar, Martian, and future missions to the outer planets* (pp. 217–237). Springer.

Sparrow, R. 2022. Human germline genome editing: On the nature of our reasons to genome edit. *American journal of bioethics* 22 (9): 4–15.

Sreedhar, S. 2017. Feminist history of philosophy. In C. Hay (Ed.), *Philosophy: Feminism* (pp. 141–166). Macmillan Reference USA.

Stoljar, N., and C. Mackenzie. 2022. Relational autonomy in feminist bioethics. In W. A. Rogers, J. L. Scully, S. M. Carter, V. A. Entwistle, and C. Mills (Eds.), *The Routledge handbook of feminist bioethics* (pp. 71–83). Routledge, Taylor & Francis Group.

Stone, A. 2004. On the genealogy of women: A defence of anti-essentialism. In S. Gillis, G. Howie, and R. Munford (Eds.), *Third wave feminism: A critical exploration* (pp. 85–96). Palgrave Macmillan.

Stone, A. 2007. *An introduction to feminist philosophy.* Polity.

Stone, A. 2019. *Being born: Birth and philosophy.* Oxford University Press.

Sturgill, L. A., S. G. Shields, and L. M. Candib. 2019. Technology and the ethical practice of reproductive care: A woman-centered lens. In L. d'Agincourt-Canning and C. Ells (Eds.), *Ethical issues in women's healthcare: Practice and policy* (pp. 233–257). Oxford University Press.

Superson, A. M. 2009. *The moral skeptic.* Oxford University Press.

Superson, A. M. 2014. The right to bodily autonomy and the abortion controversy. In A. Veltman and M. Piper (Eds.), *Autonomy, oppression, and gender* (pp. 301–325). Oxford University Press.

Swinth, K. 2018. *Feminism's forgotten fight: The unfinished struggle for work and family.* Harvard University Press.

Szocik, K. 2021. Humanity should colonize space in order to survive but not with embryo space colonization. *International journal of astrobiology* 20: 319–322.

Szocik, K. 2022. The "staying alive" theory reinforces stereotypes and shows women's lower quality of life. *Behavioral and brain sciences* 45: E146. https://doi.org/10.1017/S0140525X2 2000395.

Szocik, K. 2023a. *The bioethics of space exploration.* Oxford University Press.

Szocik, K. 2023b. Cognitive enhancement inevitably leads to discrimination against women. *AJOB neuroscience* 14 (4): 357–359.

Szocik, K. 2023c. The evolution of puritanical morality has not always served to strengthen cooperation, but to reinforce male dominance and exclude women. *Behavioral and brain sciences* 46: e316. doi:10.1017/S0140525X23000523.

Szocik, K. 2023d. Why moral bioenhancement in future space missions may not be a good idea: The perspective of feminist bioethics of space exploration. *Technology in society* 75: 102365.

Szocik, K. 2024. Proxy failure in social policies as one of the main causes of persistent sexism and racism. *Behavioral and brain sciences.*

Szocik, K., and M. J. Reiss. 2023. Why space exploitation may provide sustainable development: Climate ethics and the human future as a multi-planetary species. *Futures* 147: 103110.

Szocik, K., and M. Häyry. 2024a. Why it is rational to expect the horrible—The future of humanity and climate change. *South African Journal of Philosophy* 43 (1): 12–20.

Szocik, K., and M. Häyry. 2024b. Climate change and anti-natalism: Between the horrible and the unthinkable. *South African Journal of Philosophy* 43 (1): 21–29.

Tännsjö, T. 1998. *Hedonistic utilitarianism*. Edinburgh University Press.

ten Have, Henk A. M. J. 2022. *Bizarre bioethics: Ghosts, monsters, and pilgrims*. Johns Hopkins University Press.

Teodorescu, A. 2018. The women-nature connection as a key element in the social construction of Western contemporary motherhood. In D. A. Vakoch and S. Mickey (Eds.), *Women and nature? Beyond dualism in gender, body, and environment* (pp. 77–95). Routledge, Taylor & Francis Group.

Thone, R. R. 1997. *Fat—a fate worse than death? Women, weight, and appearance*. Haworth Press.

Tomlinson, B. 2019. *Undermining intersectionality: The perils of powerblind feminism*. Temple University Press.

Tomm, W., and G. Hamilton (Eds.). 1988. *Gender bias in scholarship: The pervasive prejudice*. Wilfrid Laurier University Press for the Calgary Institute for the Humanities.

Tong, R. 1993. *Feminine and feminist ethics*. Wadsworth.

Tong, R. 1996. Feminist approaches to bioethics. In S. M. Wolf (Ed.), *Feminism & bioethics: Beyond reproduction* (pp. 67–94). Oxford University Press.

Tong, R. 1997. *Feminist approaches to bioethics: Theoretical reflections and practical applications*. Westview Press.

Tong, R. 1998. Feminist ethics. In R. Chadwick (Ed.), *Encyclopedia of applied ethics*, vol. 2 (pp. 261–268). Academic Press.

Tong, R. 2004. Feminist perspectives, global bioethics, and the need for moral language translation skills. In R. Tong, A. Donchin, and S. Dodds (Eds.), *Linking visions: Feminist bioethics, human rights, and the developing world* (pp. 89–104). Rowman & Littlefield.

Tong, R. 2009. *Feminist thought: A more comprehensive introduction*. Westview Press.

Tong, R. 2022. Towards a feminist global ethics. *Global bioethics* 33 (1): 14–31.

Toole, B. 2019. From standpoint epistemology to epistemic oppression. *Hypatia* 34 (4): 598–618.

Toole, B. 2021. Recent work in standpoint epistemology. *Analysis* 81 (2): 338–350.

Torres, P. 2018. Space colonization and suffering risks: Reassessing the "maxipok rule." *Futures* 100: 74–85.

Traphagan, J. W. 2019. Which humanity would space colonization save? *Futures* 110: 47–49.

Traphagan, J. W. 2020. Religion, science, and space exploration from a non-Western perspective. *Religions* 11 (8): 397.

Tremain, S. L. 2019. Feminist philosophy of disability: A genealogical intervention. *Southern journal of philosophy* 57: 132–158.

Tuana, N. 2023. *Racial climates, ecological indifference: An ecointersectional analysis*. Oxford University Press.

Turkmendag, I. 2022. Exploitation and control of women's reproductive bodies. In W. A. Rogers, J. L. Scully, S. M. Carter, V. A. Entwistle, and C. Mills (Eds.), *The Routledge handbook of feminist bioethics* (pp. 486–500). Routledge, Taylor & Francis Group.

Tyszczuk, R. 2021. Collective scenarios: Speculative improvisations for the Anthropocene. *Futures* 134: 102854. https://doi.org/10.1016/j.futures.2021.102854.

Van Slyke, J. A., and K. Szocik. 2020. Sexual selection and religion: Can the evolution of religion be explained in terms of mating strategies? *Archive for the psychology of religion* 42 (1): 123–141.

Vandermassen, G. 2005. *Who's afraid of Charles Darwin? Debating feminism and evolutionary theory*. Rowman & Littlefield.

Vandermassen, G. 2008. Can Darwinian feminism save female autonomy and leadership in egalitarian society? *Sex roles* 59: 482–491.

Vaughn, L. 2023. *Bioethics: Principles, issues, and cases.* Oxford University Press.

Vint, S. 2021. *Biopolitical futures in twenty-first-century speculative fiction.* Cambridge University Press.

Von Morstein, P. 1988. Epistemology and women in philosophy: Feminism is a humanism. In W. Tomm and G. Hamilton (Eds.), *Gender bias in scholarship: The pervasive prejudice* (pp. 147–164). Wilfrid Laurier University Press for the Calgary Institute for the Humanities.

Walters, M. 2005. *Feminism: A very short introduction.* Oxford University Press.

Waring, S. P., and B. C. Odom (Eds.). 2019. *NASA and the long civil rights movement.* University Press of Florida.

Warren, K. J. 2015. Feminist environmental philosophy. In E. N. Zalta (Ed.), *The Stanford encyclopedia of philosophy,* Summer 2015 ed. https://plato.stanford.edu/archives/sum2015/entries/feminism-environmental/.

Watson, L. 2017. Feminist perspectives on human nature. In C. Hay (Ed.), *Philosophy: Feminism* (pp. 71–99). Macmillan Reference USA.

Watson, L., and C. Hartley. 2018. *Equal citizenship and public reason: A feminist political liberalism.* Oxford University Press.

Weaver, S., and C. Fehr. 2017. Values, practices, and metaphysical assumptions in the biological sciences. In A. Garry, S. J. Khader, and A. Stone (Eds.), *The Routledge companion to feminist philosophy* (pp. 314–327). Routledge, Taylor & Francis Group.

Weitekamp, M. A. 2004. *Right stuff, wrong sex: America's first women in space program.* Johns Hopkins University Press.

Wells-Jensen, S. 2022. Welcoming disability as necessary in space travel. In C. S. Cockell (Ed.), *The institutions of extraterrestrial liberty* (pp. 475–482). Oxford University Press.

Wells-Jensen, S. 2023. Occupy space: Will disabled people fly? In J. S. J. Schwartz, L. Billings, and E. Nesvold (Eds.), *Reclaiming space: Progressive and multicultural visions of space exploration* (pp. 232–240). Oxford University Press.

Wells-Jensen, S., J. A. Miele, and B. Bohney. 2019. An alternate vision for colonization. *Futures* 110: 50–53.

Whipps, J. D. 2017. A historical introduction: The three waves of feminism. In C. Hay (Ed.), *Philosophy: Feminism* (pp. 3–33). Macmillan Reference USA.

Whitman Cobb, W. N. 2022. Stubborn stereotypes: Exploring the gender gap in support for space. *Space policy* 54, 101390.

Whitman Cobb, W. N. 2024. For all (wo)mankind: Advancing a feminist critique of US space policy. *Space policy* 67: 101594.

Wieczorek, P. 2023. *Imagining the anthropocene future: Body and the environment in indigenous speculative fiction.* Peter Lang.

Wilkerson, I. 2020. *Caste: The origins of our discontents.* Random House.

Wilson, E. A. 2004. *Psychosomatic: Feminism and the neurological body.* Duke University Press.

Wilson, Y., A. White, A. Jefferson, M. Danis. 2019. Intersectionality in clinical medicine: The need for a conceptual framework. *American Journal of Bioethics* 19 (2): 8–19.

Wing, A. (Ed.). 2003. *Critical race feminism: A reader.* 2nd ed. New York University Press.

Withers, A. J., L. Ben-Moshe (Eds.) with L. X. Z. Brown, L. Erickson, R. da Silva Gorman, T. A. Lewis, L. McLeod, and M. Mingus. 2019. Radical disability politics. In R. Kinna and U. Gordon (Eds.), *Routledge handbook of radical politics* (pp. 178–193). Routledge.

Witt, C. 2011. *The metaphysics of gender.* Oxford University Press.

Wolf, N. 1992. *The beauty myth: How images of beauty are used against women.* Anchor Books.

Wolf, S. M. 1996. Introduction: Gender and feminism in bioethics. In S. M. Wolf (Ed.), *Feminism & bioethics: Beyond reproduction* (pp. 3–43). Oxford University Press.

World Commission on Environment and Development. 1987. *Our common future.* Oxford University Press.

Wright, D. W. M. 2016. Hunting humans: A future for tourism in 2200. *Futures* 78–79: 34–46.

Young, I. M. 2005. *On female body experience: "Throwing like a girl" and other essays.* Oxford University Press.

Zack, N. (Ed.). 2017. *The Oxford handbook of philosophy and race.* Oxford University Press.

Zinn, M. B., and B. T. Dill. 1994. Difference and domination. In M. B. Zinn and B. T. Dill (Eds.), *Women of color in U.S. society* (pp. 3–12). Temple University Press.

Index